Non-Volatile CBRAM/MIM Switching Technology for Electronically Reconfigurable Passive Microwave Devices

"…ajnaana-timirandhasya jnanaanjana-salakaya caksur-unmilitam yena tasmai sri-guruve namah…"

(Excerpts from Indian philosophy in Sanskrit language)

I humbly express my deep gratitude and obeisance to all my dear teachers who opened my eyes to the wisdom of knowledge, away from the blindness of ignorance, in all phases of my life…

(Translation)

Series Editor
Etienne Perret

Non-Volatile CBRAM/MIM Switching Technology for Electronically Reconfigurable Passive Microwave Devices

Theory and Methods for Application in Rewritable Chipless RFID

Jayakrishnan Methapettyparambu Purushothama
Etienne Perret
Arnaud Vena

iSTE

WILEY

First published 2022 in Great Britain and the United States by ISTE Ltd and John Wiley & Sons, Inc.

ISTE Ltd
27-37 St George's Road
London SW19 4EU
UK

www.iste.co.uk

John Wiley & Sons, Inc.
111 River Street
Hoboken, NJ 07030
USA

www.wiley.com

Library of Congress Control Number: 2021949301

British Library Cataloguing-in-Publication Data
A CIP record for this book is available from the British Library
ISBN 978-1-78630-813-9

MIX
Paper from responsible sources
FSC® C013604

Contents

Preface

> Science means constantly walking a tightrope between blind faith and curiosity; between expertise and creativity; between bias and openness; between experience and epiphany; between ambition and passion; and between arrogance and conviction – in short, between an old today and a new tomorrow[1].
>
> Heinrich Rohrer
> (Swiss Physicist)

Radio has made people smarter since its discovery in the late 1800s. Now scientists are trying to make radios smart to make life even easier. Electromagnetic propagating waves in the range of 30 kHz–300 GHz are called radio waves, and electrical equipment handling these propagations are generally called radio or radio frequency (RF) devices. Human life is dependent on RF electronics in our everyday routine. TV and radio broadcast, mobile phone communication, the Internet, microwave heating and so on are an integral part of our daily life, implicitly or explicitly.

Radio equipment can be an active device (that uses an active power source for operation), like a mobile phone or a television set, or passive equipment (that does not use an active power source for operation or data storage, but scavenges a part of an interrogating energy source), like a contactless credit card, or an anti-theft RFID label in a supermarket.

---

1 Heinrich Rohrer Quotes. (n.d.). BrainyQuote.com. Retrieved July 9, 2020, from BrainyQuote.com web site: https://www.brainyquote.com/quotes/heinrich_rohrer_736250.

Communication, connectivity and surveillance are the major fields that utilize most of the RF electronics in this era. This may involve the connectivity between two individuals separated in two continents or between two electrical devices in a same drawing room. The advent of the Internet of things (IoT), backed by 5G communication technologies, has intensified this demand by providing connectivity to almost all our daily utility accompaniments. However, all this electrical equipment, whether active or passive, requires electrical switches, beginning with the initial need for power, and these span the entirety of the internal RF circuitry. These switches operate in different domains, ranging from electrical power switches to different relays, selectors, limit switches and RF switches.

Among these, RF switches play a key role in their domain and are unique with respect to required performances. These are used not only in wireless communications, but also in smart RF sensors and actuators of the future. RF switches allow electronic reconfiguration of device parameters such as operating frequency in a filter, radiation pattern in an antenna and so on, in order to make these devices more efficient. They are present in all communication systems and allow, in summary, signal multiplexing or regulation, so as to reconfigure, or dynamically control, the system. Today's technological solutions for RF switching (solid-state semiconductor switches and RF micro-electromechanical-systems [MEMS]) need some improvements to meet the emerging requirements, in which the most desired innovation is non-volatile operation, i.e. operating without any energy requirements for maintaining an impedance state. Such a broadband solution, based on a flexible and low cost approach, is eagerly awaited.

Conductive bridging random access memory/metal insulator metal (CBRAM/MIM) switches are a new innovation of memory technology, which has also been identified as a potential non-volatile RF switching solution in the recent years. The primary target of this book, and the core concept of the studies presented herewith, is to introduce the idea of a new generation of electronically reconfigurable, solid-state and passive RF-microwave devices based on integrated CBRAM/MIM switching technology. We will present our work and results, focusing on the development of a new technology of low-cost non-volatile RF switches with both high performance and flexibility of implementation.

We demonstrate the basic principle of CBRAM/MIM RF switches and give proof of concept realizations of integration of this technology into passive RF and microwave devices, such as electronically rewritable chipless RFID tags, which are often referred to as the "rewritable RF barcodes of the future", electronically pattern steerable antennas, electronically reconfigurable filters and single pole double throw (SPDT) switches on classic as well as flexible substrates. The necessary analysis and theoretical validations of these concepts are also presented with these results.

We take this work as a humble beginning for a new innovation of electronically reconfigurable and non-volatile passive RF and microwave devices, which is dedicated to the future, and nurtured from deeply within our heart, with our passion and wide-eyed dreams for the betterment of current science and technology. Our investigations and findings are organized into four chapters in this book, as described below.

In Chapter 1, a general idea of the motivation for and background of the need for a non-volatile RF switch is discussed in detail. A review of prominent RF switching technologies is presented, along with the state of the art of research in CBRAM/MIM RF switch technology. The requirement of non-volatile RF switches and the reasons for the choice of CBRAM/MIM technology for this are also explained in detail.

Chapter 2 states the real-world implementation challenges of a low-cost non-volatile RF switching technology and describes our efforts to overcome these limitations. We present in this chapter the design and optimization process of CBRAM/MIM based Coplanar waveguide (CPW) RF switches on classic as well as flexible paper substrates, using simple and low-cost methods, compatible with mass industrial production. Discussion of electrical equivalent models, time stability and other interesting features are also given in addition in this chapter. An application example of these switches in SPDT configuration is also included in this chapter.

Chapter 3 focuses on the design and development of solid-state electronically rewritable chipless RFID tags, which may be informally called "rewritable RF barcodes". In this chapter, we describe in brief the principle of chipless RFID and present the design and realization of electronically rewritable chipless RFID tags on classic and flexible substrates. Electrical equivalent model analysis and advantages of using the proposed integrated CBRAM/MIM switches for chipless RFID applications are discussed in detail.

Chapter 4 explains the application of the proposed CBRAM/MIM RF switch technology to electronically reconfigurable filter and antenna models using simple and low-cost methods. We present the implementation of reconfigurable band-stop filters, backed by theoretical discussions and electrical model analysis. We also discuss the extension of the proposed techniques to other filter topologies and the mitigation of current limitations of realized filters in futuristic designs. Application of CBRAM/MIM RF switches for electronically reconfigurable antennas for radiation pattern steering is also presented affirmatively in this chapter. We conclude this chapter by discussing the advantage of using the proposed RF switch technology for reconfigurable antenna and filter applications.

This book also has an appendix, which contains some interesting outcomes of a control experiment for observing the filament formation in CBRAM/MIM switches.

Jayakrishnan METHAPETTYPARAMBU PURUSHOTHAMA
Etienne PERRET
Arnaud VENA
November 2021

1

Motivation and Background: RF Switches and the Need for a Non-Volatile RF Switch

1.1. Introduction

In electrical engineering, a switch, in general, is a device which opens, closes or regulates a current path at a desired event, like ON or OFF. Switches are an inevitable component of all electronic systems we come across in our life. Like the presence of salt in food, most of the time we are unaware of the presence of this mighty little electrical component that plays many roles in our daily life.

Even though the basic responsibility of a switch is to control (open or close or regulate) a current path when desired, it can perform multiple functions, such as a classic vintage mechanical electrical switch, a telephone keypad, a relay in an automatic bread toaster, a switching regulator in a room heater, a capacitive or resistive touch screen of a mobile phone, an integral part of signal modulation system for communication and Internet and so on.

In a broad sense, an electrical switch is identified by the associated actuation mechanism used to close, open, or regulate the current path through it. Electrical switching action is usually associated with the following actuation mechanisms: mechanical (mechanical toggle switches), electronic (PIN diodes, field effect transistor [FET], electromechanical relays), chemical (azobenzene photoelectric switches), magnetic (reed switches), light (opto-couplers), software defined (routers), ionic bridging (conductive bridging random access memory [CBRAM]), phase change and so on. This definition, however, is a very crude explanation of available switch technologies, which are in greater abundance than could be analyzed or differentiated in a single book.

In this book, we concentrate on non-volatile (which do not require a power supply to maintain their impedance state) switches for radiofrequency (RF) and microwave frequencies using nanoionic conductive-bridging metal–insulator–metal (MIM) switch technology, generally identified as CBRAM, for low-cost RF and microwave devices. Here, we focus on overcoming one of the significant shortcomings of passive RF and microwave devices, which is non-volatile electronic reconfigurability.

In this book, we propose techniques to make possible the integration of passive non-volatile switches with passive devices that primarily work in the frequency range of a few MHz to around 10 GHz. Referring to this frequency range, we assume the freedom to address the domain of this book as "RF" or "microwave" frequencies, interchangeably. This book aims to democratize the CBRAM/MIM switching technology for passive/low-power RF and microwave devices using a simple manufacturing process, which is compatible with mass production, at an economically efficient budget.

In this chapter, we discuss and give some examples about the need for such a study, and the background information available at the start of the research. We give a brief introduction on the available RF and microwave switch technologies. Then we present the state of the art in memristive CBRAM/MIM switch technology, and explain the reasons for choice of this technology in this research. We also provide in brief an introduction to targeted RF domains, which are the point of application of the developed CBRAM/MIM RF switching solutions.

1.2. Requirements and definition of a switch at RF and microwave frequencies

Modern life is predominantly dependent on electromagnetic energy propagations in the range 3 kHz–300 GHz on the electromagnetic spectrum, which is generally referred to as RF and microwave signals (Pozar 2005). The RF and microwave spectrum includes numerous application bands such as television, radio, mobile phone, GPS, Bluetooth, Wi-Fi, satellite and so on. RF and microwave radiations are also used for a lot of extended applications such as tracking, ranging and enhanced vision (radar and radio astronomy), heating (microwave ovens and heating in nuclear fusion reactors), imaging (microwave tomography), medical applications (microwave ablation of tissues) and so on. In essence, one could say that RF and microwave frequency signals find their application anywhere from a simple domestic kitchen to outer space.

High-frequency RF and especially microwave signals have very short free space wavelengths, typically of the order of a few tens to a few centimeters. Due to high frequency and small wavelengths, standard circuit theory and electronic circuit approaches are not enough to solve and define working phenomena in this domain (Jordan and Balmain 1968; Bahl 2003; Pozar 2005; Sadiku 2007). At these frequencies, classic lumped circuit element approximations defined by standard circuit theory prove difficult to hold onto, as at RF and microwave frequencies components are often associated with a number of distributed parameter effects arising from their geometry. This is because at these frequencies the phase of a voltage and current is significantly affected by the physical size of a device due to small wavelengths. Similarly, circuit parameters, such as inductance, capacitance and associated dielectric properties, are significant for circuit element sizes comparable to wavelengths. At these wavelengths and frequencies, electromagnetic field theory and broadly defined circuit theory, both described by Maxwell's equations, are used for the analysis (Jordan and Balmain 1968; Bahl 2003; Pozar 2005; Sadiku 2007).

Similar to standard electronic circuits, switches are an integral and also inevitable part of RF and microwave circuits. However, some special care should be taken and some requirements should be satisfied for the design, development and applications of switches in this field. This is with respect to the previous explanations.

The performance and behavioral features desired for RF and microwave switches are not much different from those expected for any devices or circuit at these frequencies (Breed 2010). The most relevant of these requirements or performance factors are briefed as follows:

– bandwidth of operation of the switch;

– ON and OFF state insertion loss and isolation at frequencies of interest;

– switching speed;

– operating lifetime (number of switching cycles and reliability);

– figure of merit (product of OFF state capacitance and ON state impedance of the switch, typically expressed in THz);

– operating voltage and power consumption;

– power handling, stability and linearity;

– temperature and environmental dependence;

– isolation of biasing path from RF signal path;

– cross-talk and radiation isolation (isolation to other components in the circuitry).

These requirements vary according to a desired application, and are seldom obtained altogether in a single technology, but are a compromise defined with respect to a certain tradeoff, taking into account factors such as desired performance, accuracy requirements, cost, available physical space and so on.

This section and eventually this chapter chalks a boundary, helping our readers to understand and assess the outcome of the experimental results and associated discussions on CBRAM-based RF switching solutions presented herewith.

1.3. Review of RF and microwave switching technologies

A handful of switching technologies are available for RF and microwave switching, most of them with a unique distinctive feature. Some of the common technologies include PIN junction diode switches (Keysight Technologies 2019), FET switches (Keysight Technologies 2017a), hybrid switches like microwave monolithic integrated circuits (MMIC) (Mizutani and Takayama 2000; Keysight Technologies 2014), electromechanical RF switches like micro-electromechanical-systems (MEMS) (Analog Devices 2018) and so on.

New innovations such as memristive switches, including phase change memory (PCM) (Wang *et al.* 2014) or CBRAM (Pi *et al.* 2015), and engineered material switches, such as monolayer grapheme-based switches (Li and Cui 2015; Moldovan *et al.* 2015; Ma *et al.* 2016; Pan *et al.* 2017), are also in a budding stage to redefine the future of RF switching.

In this chapter, we discuss only the significant switching technologies of the above list. We also outline the state of the art of the topic in focus in this book, which is the memristive CBRAM/MIM switch technology. Here, we try to differentiate and compare this technology with the former counterparts.

Classically, RF and microwave switches could be broadly grouped into three main categories as follows:

1) electromechanical switches;

2) solid-state (semiconductor-based) switches;

3) memristive switches.

Electromechanical switches are similar to electrically actuated mechanical switches such as relays and were succeeded by more efficient technologies such as MEMS. Solid-state switches generally work on the basis of regulated electron flow through semiconductor or engineered materials, such as PIN diodes, FETs and hybrid semiconductor switches.

Technology (Right) Parameters (Down)	**Solid-state semiconductors**	**MEMS**	**Memristive switches**
ON state impedance	~1 Ω	~0.5–2 Ω	~2–10 Ω
Lower frequency limit	DC to a few kHz	DC	DC
Upper frequency limit	>100 GHz	>1 THz	>40 GHz
Switching time	10–100 ns	1–300 μs	1–10 μs
Figure of merit (THz)	1–4	10–300	0.015–38
Non-volatility	No	No	Yes
Fabrication complexity	High	Very high	High, low*
Commercial availability	Yes	Yes	No
Cost	Low	High	No commercial products
*High for classic fabrication techniques (Nessel and Lee 2011; Pi *et al.* 2015) and low for fabrication with polymer ion-conductors (Perret *et al.* 2015; Jian *et al.* 2017).			

Table 1.1. *Comparison of general categories of RF switching technologies*

Here, we take the privilege of introducing a third and emerging category called "memristive switches", which is the focus topic of this book. We present this topic as a separate category for detailed discussion to highlight its relevance and unique characteristics.

A memristive switch could in a simple way be defined as a resistor with non-volatile memory, i.e. a resistor element that is capable of preserving its impedance states based on a preset former actuation, even after its withdrawal. This is achieved by techniques such as growing a metallic filament among two electrochemically asymmetric electrodes sandwiching an ion conductor layer as in CBRAM, changing the phase of a material from amorphous to crystalline that also inverses high to low impedance state as in PCM and so on. On a broader perspective, memristive switches are a subset of solid-state switches, but are considered here individually to highlight their significance and contrast from classic solid-state semiconductor switching technologies. In this section, we first we present a summary of the comparison of performance factors of all discussed RF switching technologies for a clearer understanding for the reader. This comparison is summarized from the discussion and references in the text, and is given in Table 1.1.

In the following sections, we explain in brief the main considerations, advantages and shortcomings of each of these switching technologies.

1.3.1. *Electromechanical switches: MEMS*

MEMS are a technology of fabrication of microscopic electronic devices with moving parts (Rebeiz 2004). Typically, MEMS are made up of components of size 1–100 μm, indicating the "micro" in MEMS. Generally, these devices are fabricated using micromachining technologies on substrates such as silicon, quartz, gallium arsenide and so on.

MEMS technologies were introduced in the 1970s for the realization of pressure and temperature sensors. After that, RF switches using MEMS technology were successfully presented in the late 1990s by the United States of America's Defense Advanced Research Projects Agency (DARPA) (Rebeiz 2004). MEMS RF switches resemble classic electromechanical relay switches at a micrometer scale.

Today, ample amount of research has been done and it is still continuing on MEMS technology for RF switching, and several designs have been proposed in this domain that work up to THz frequencies (Goldsmith *et al.* 1998; Mollah and Karmakar 2001; Suzuki 2001; Chan *et al.* 2003, 2007; Rebeiz 2003; Unlu and Jarrahi 2014; Oberhammer 2017). Technological competence and reliability of MEMS technology are acknowledged by the fact that MEMS-based RF switches are now being trusted for a multitude of applications including medical and military equipment, and are widely released as commercial products (e.g. Analog Devices 2018). This technology is now used for a wide spectrum of RF switching applications including phase shifters (Rebeiz *et al.* 2002), multipole switches (Lee *et al.* 2005), reconfigurable antennas (Jung *et al.* 2006; Zohur *et al.* 2013), outer space and satellite communication applications (Souchon *et al.* 2017) and so on. There are several topologies of MEMS switches, implemented in series or shunt modes. Actuation mechanisms for MEMS switches are achieved through several techniques such as electrostatic, magnetostatic, piezoelectric or thermal means (Rebeiz 2004).

Broadly, two types of MEMS switch technologies are used extensively for RF switching, namely ohmic contact type and capacitive contact type. The former technology is used for frequency ranges from DC to 60 GHz and the later for 6–200 GHz, in general (Rebeiz 2004). These topologies are both implemented in series and shunt mode.

In terms of ohmic contact MEMS switches, the device consists of a micro-machined moving electrical contact arm, which is actuated to make or break an ohmic electrical contact (e.g. Chan *et al.* 2007). Figure 1.1 shows an illustration of an ohmic contact MEMS RF switch in series configuration. Here, the cantilever signal bridge (connected to signal line) is actuated to make or release an ohmic electrical contact with the isolated transmission line, respectively, for the ON and OFF states.

For capacitive contact MEMS switches, a variable capacitance contact system is made with the help of a suspended/spanning membrane to couple RF signal energy to a required transmission path, or to ground, for series and shunt configurations, respectively, through a capacitive contact (Goldsmith *et al.* 1998). Figure 1.2 shows an illustration of the working of a shunt mode capacitive MEMS RF switch built in a CPW transmission line configuration. Here, the suspended membrane bridge is actuated as shown in figure to touch and release from the dielectric slab to exhibit high and low capacitance values, respectively, to short circuit (OFF state) the power flowing through transmission line to ground for high capacitance values, and to revert (ON state) the signal flow in transmission line for low value. Capacitive contact MEMS switches are generally observed to exhibit more reliability than ohmic contact types (Rebeiz 2004). Ohmic contact type switches are prone to issues such as contact conductance degradation due to friction and oxide formations, cold welding at atomic level in metals leading to failure of switches and so on.

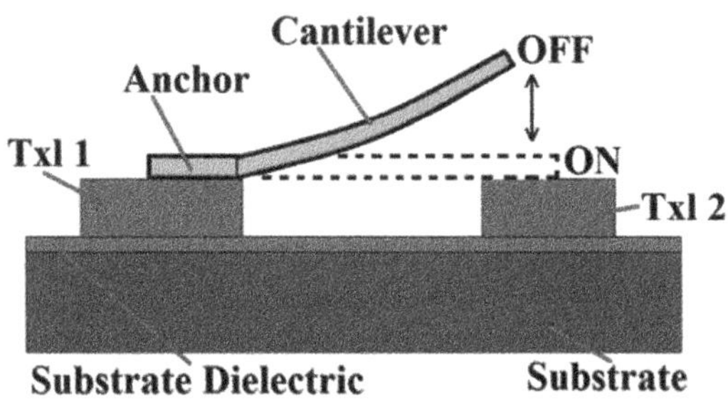

Figure 1.1. *Ohmic contact type MEMS RF switch in series configuration*

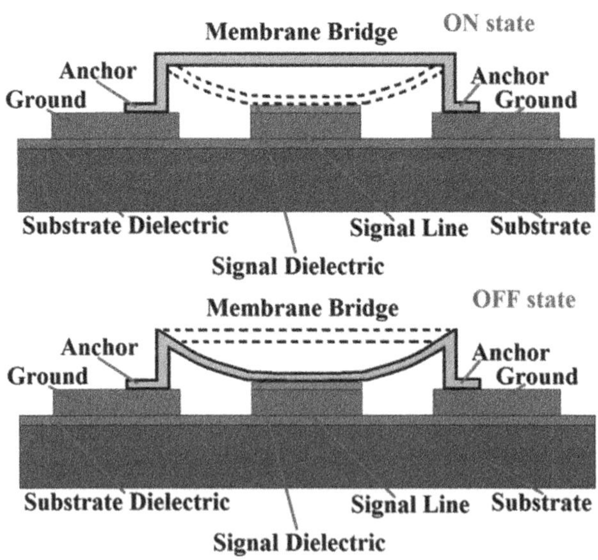

Figure 1.2. *Capacitive contact type MEMS RF switch in shunt configuration on a CPW line, showing ON and OFF states (cross-section view). For a color version of this figure, see www.iste.co.uk/jayakrishnan/technology.zip*

MEMS switches are known for their superior performance in terms of low insertion loss, high isolation and high linearity in comparison to solid-state semiconductor switches. However, MEMS switches are limited by factors including fabrication cost, size and robustness related to the reliability of moving parts, with respect to solid-state semiconductor counterparts.

1.3.2. *Solid-state semiconductor switches*

Solid-state semiconductor switches are based on regulation of RF energy flow through a semiconductor material (generally unheated) by controlling its electrical conductivity. This is done by introducing or inhibiting charge carriers (electrons or holes) in the semiconductor material by means of an external DC bias. The word "solid state" was introduced around 1960 and means that the switch involves only unheated solid semiconductor materials and no moving parts, as in the immediate counterparts like mechanical relays, nor vacuum tubes, which are now obsolete anyway. In this book, we provide detailed discussion only on solid-state semiconductor switches limited to RF and microwave applications. The most common and extensively used solid-state semiconductor RF switches are PIN diodes, FETs and hybrid switches, discussed in detail below.

Generally, solid-state semiconductor RF switches show more reliability and lifetime characteristics in comparison to MEMS RF switches or their ancient counterparts, the vacuum tubes. This is due to their remarkable resistance to shocks, vibrations and mechanical wear (Keysight Technologies 2017b). They also exhibit a faster switching time and are cheaper in cost, in comparison to MEMS. The adaptiveness of semiconductor devices to be batch produced in groups of thousands or more on a single wafer is one of the major reasons for their reduced cost. One of the drawbacks of semiconductor RF switches compared to MEMS is their higher ON state impedance, which leads to higher insertion losses.

1.3.2.1. *PIN diode RF switches*

A PIN diode is basically a semiconductor junction diode with layer architecture as indicated by its name. It has a high resistivity intrinsic (I) semiconductor layer sandwiched among a pair of highly doped P-type (P+) and N-type (N+) semiconductor layers as shown in Figure 1.3(a).

When this diode is forward biased, it starts functioning similar to a regular P-N junction diode. But due to the presence of the high resistivity intrinsic layer, charge carriers from their respective layers (electrons from N type and holes from P type) are injected into the intrinsic (I) layer and recombine there. However, due to the high doping concentration of P and N layers, and high resistivity of I layer, the charge injection rate is much higher than the recombination rate. This results in an accumulation of many charge carriers in the I layer despite recombination and leads to effective lowering of resistance of this layer. This establishes the RF-ON state of the PIN diode switch. During forward bias, a PIN diode could be represented as an RL network where the resistance R_F is the resistance of the diode under forward bias, and L_P is the parasitic inductance due to the associated package, as shown in Figure 1.3(b).

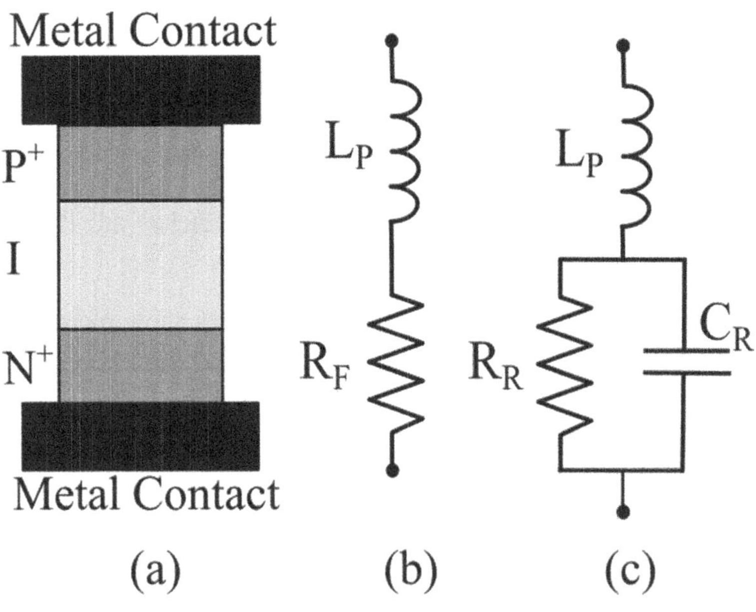

Figure 1.3. *PIN diode: (a) layer architecture of a typical PIN diode; (b) electrical equivalent circuit in forward bias (ON state); (c) electrical equivalent circuit in reverse bias (OFF state)*

During reverse bias (or no bias), the intrinsic region does not have any charge carriers and exhibit high resistivity, establishing an RF-OFF state of the diode switch. At this state, the electrical model of the diode could be represented as an RC parallel network in series with the parasitic inductance L_P, as shown in Figure 1.3(c). Here, R_R is the resistance measured across the diode in reverse bias and C_R is the diode junction capacitance. This junction capacitance is an important feature as it limits the maximum frequency of operation of the PIN diode switch. Nevertheless, typical commercial PIN diodes have usable frequency range up to 50 GHz (Keysight Technologies 2017b). The dielectric relaxation frequency of intrinsic (I) regions sets the lower frequency limit of a PIN diode, the typical values of which are around a few MHz. PIN diodes are basically current controlled devices, and in terms of power consumption, PIN diodes consume higher power than their immediate relatives, the FET switches. Another important characteristic of PIN diode RF switches is that in these devices, the DC biasing path is not different from the RF path, and hence special isolation circuits are necessary to prevent leakage and unwanted mixing of both these paths.

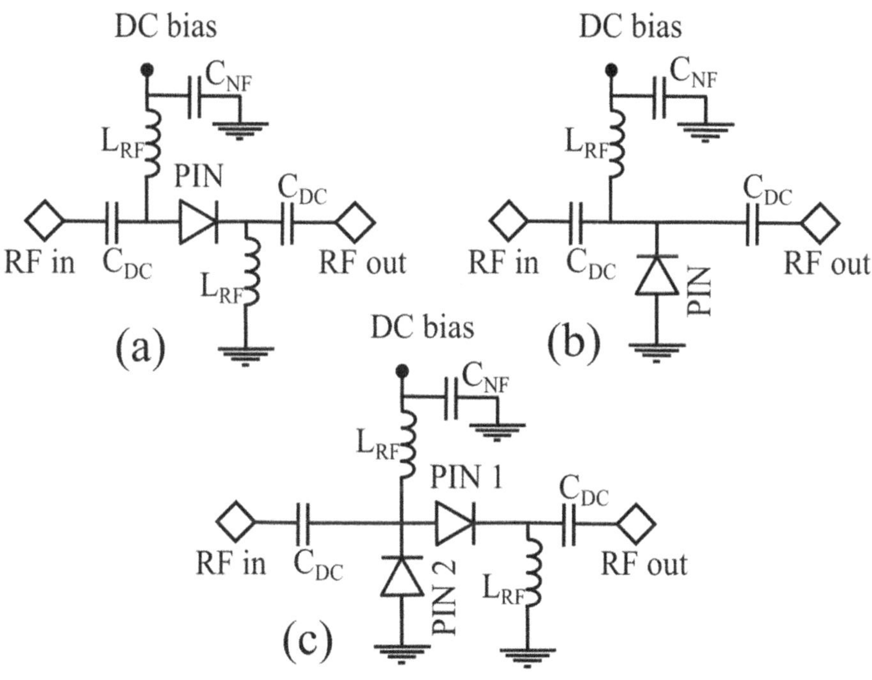

Figure 1.4. *Application of PIN diodes as switches in RF circuit: (a) series mode, (b) shunt mode and (c) series–shunt mode*

PIN diodes are used in circuits as RF switches in series, shunt and as a combination of series–shunt configurations, as shown in Figure 1.4, according to the application and performance requirements. Here, C_{DC} is the DC block capacitor to prevent mixing of DC biasing used for ON and OFF control of PIN diodes into the RF path, L_{RF} is the RF choke to prevent the leakage of RF power into the DC biasing path and C_{NF} is the noise bypass capacitor to discharge any unwanted noise pickup from the DC side of the circuit. Among these components, RF choke inductors are a factor that add significantly to the implementation cost of this circuit.

1.3.2.2. *FET-based RF switch*

FET are semiconductor devices in which the current flow in a heavily doped two terminal (source, drain) unipolar semiconductor block is regulated based on an applied electric field to an attached semiconductor layer (gate) of opposite polarity (Keysight Technologies 2017b). FETs are introduced with different layer architectures and technologies, depending on materials used, electrical contact among gate to drain and source, construction topology and so on; however, in most cases, the principle remains the same. Some of the notable models that are used commercially are MOSFET (metal oxide semiconductor field effect transistors), MESFET (metal semiconductor field effect transistors) and HFET (heterostructure field effect transistors) (Storm 1967).

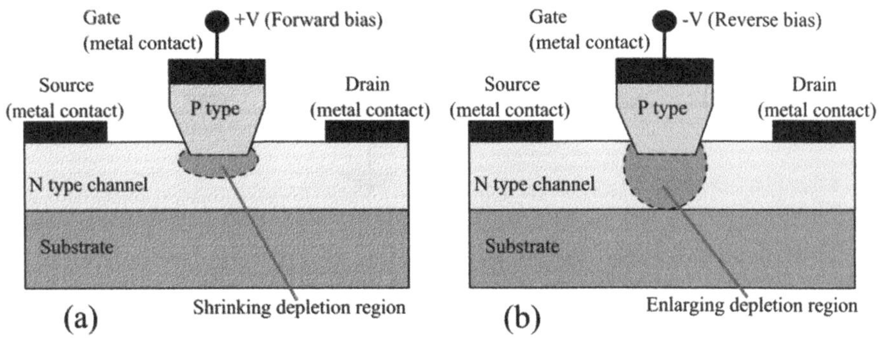

Figure 1.5. *Simplified layer architecture of an N-channel field effect transistor (FET) for RF switching applications. (a) ON state and (b) OFF state*

Similar to a PIN diode, a specially constructed FET (taking into account the requirements stated in section 1.2) could be used as an RF switch (Keysight Technologies 2014, 2017a, 2017b). Figure 1.5 shows a simplified layer diagram of an N-channel Gallium arsenide (GaAs) FET for RF application.

Here, the drain to source path offers a low resistance current path to an incoming RF signal during zero or forward bias, establishing the ON state of the switch, as shown in Figure 1.5(a). Inversely, one could see that a reverse bias voltage applied to the P-type gate material and one of the terminals (source or drain) enables the increase of the drain to source resistance by enlarging the highly resistive depletion region, establishing the OFF state of the switch as shown in Figure 1.5(b). Depletion of the conductive channel in this manner is referred to as a "pinch off" mechanism. This mechanism provides excellent isolation properties even at low frequencies and DC for an FET RF switch. The DC biasing path is different from the RF signal path for an FET, and this helps in reducing the circuit complexity by avoiding the use of comparatively expensive RF choke inductors and DC blocks.

FET is a voltage-controlled device, and it is characterized by low power consumption in operation, in comparison to a PIN diode switch. However, the ON state channel resistance of an FET is typically greater than that of a PIN diode, which results in higher insertion loss in comparison to the latter one. The operating frequency range of FET switches are also inferior to PIN diodes due to the presence of greater drain to source capacitance in reverse bias (OFF state), as a result of thinner depletion region. Figure 1.6 shows an example of a classic configuration of a series–shunt FET RF switch in an SPDT RF switching application.

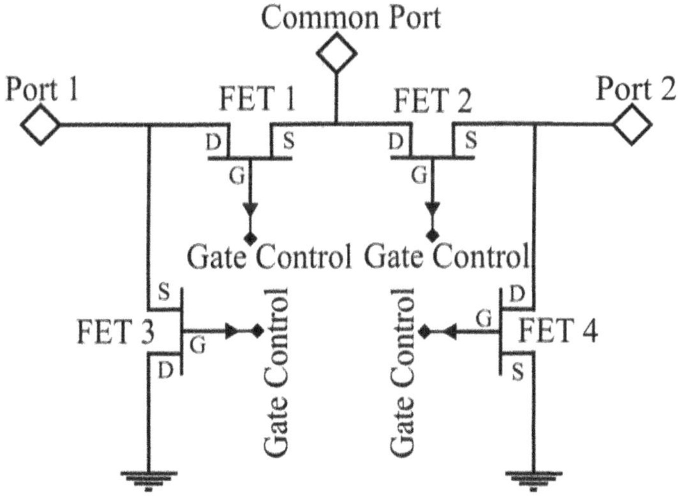

Figure 1.6. *Application of FETs as RF switches in a single pole double throw (SPDT) configuration*

We have seen the merits and demerits unique to both FET and PIN diode switches in previous sections. Engineers have combined these advantages to get best out of these technologies in the form of hybrid RF switches. Hybrid RF switches is a general nomenclature, given to a group of semiconductor switch technologies that aims at manufacturing of RF switches with optimized performance features using modified forms of PIN diodes and FETs. Extensive research has been done on this topic for a few decades, and today a number of excellent switching solutions have been reported. Some of the commonly used technologies are MMIC (Mizutani and Takayama 2000), SOI (silicon on chip) (Lee and Lee 2010; Jaffe *et al.* 2015), RF–CMOS (complementary metal–oxide–semiconductor) (Li *et al.* 2008) and so on. In commercial versions of these technologies, the switches are modular and are optimized for features such as input and output impedances, cross-talk and electromagnetic interference reduction, mechanical package stability and vulnerability to extreme environments and temperatures.

1.3.3. *Memristive RF switches*

Memristive RF switches are a new innovation among the classic and mature RF switching technologies explained above. As the name indicates, a memristor is a resistor with non-volatile memory (Chua 2011). This is a hypothetical electronic device, predicted by a circuit theorist called Leon Chua in his revolutionary article entitled "Memristor – The Missing Circuit Element" (Chua 1971). Chua predicted, based on the arguments of symmetry connecting basic electrical quantities, that there exists the possibility of a fourth fundamental element, which he named the "memristor".

The theories of the memristor concept are too formal and too far from the evolved applications of memristive RF switches. However, this book would be incomplete without stating the general principle behind the focus topic of this publication, so we will state the general theory of the memristor, and then move on to memristive RF switching technologies.

Chua's prediction of the memristor is described below as a generalized overview, and it is based on the following arguments of symmetry. The four fundamental variables of electricity and magnetism, the electric charge (q), magnetic flux (ϕ), voltage (v) and current (i), always co-exist, and these quantities except magnetic flux and charge are connected with each other by the so-called fundamental circuit elements resistance (R), inductance (L) and capacitance (C).

Chua stated that there should be a nonlinear relationship to connect the magnetic flux and charge, as shown in equation [1.1]. A circuit element called a memristor with a property called memristance (M) was modeled based on this theory, as shown in equation [1.2], which could be simplified by substituting flux as time integral of voltage and charge as time integral of current as in equation [1.3]. This concept is summarized in the form of an illustration in Figure 1.7. Figure 1.7 also shows the symbol of a memristor devised by Chua. From equation [1.3], the unit of a memristor could be extracted as the Weber per Coulomb (Wb/C) or simply Ohm (Ω).

$$f(\phi(t), q(t)) = 0 \qquad [1.1]$$

$$M(q) = \frac{d\phi}{dq} \qquad [1.2]$$

$$M(q(t)) = \frac{d\phi/dt}{dq/dt} = \frac{v(t)}{i(t)} \qquad [1.3]$$

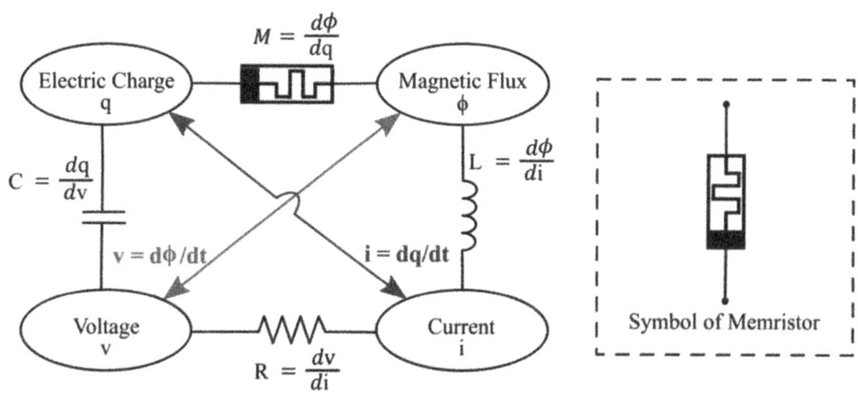

Figure 1.7. *Concept of relation among electric charge (q), magnetic flux (ϕ), voltage (v) and current (i) in terms of fundamental circuit elements: resistance (R), inductance (L) and capacitance (C), and the proposed memristor (M). Inset shows electrical symbol of a memristor*

Summing up the above theory, what is interesting for us is the concept of the memristor, which could be stated in a simplified way as follows. The memristor is an electrical resistance component/parameter with the following property: its resistance is not constant, but depends on the history or memory of previous current flow through this device (Chua 1971; Strukov *et al.* 2008). This makes the resistance of a memristor a function of past recollection of current flow, or amount of magnetic flux linked per unit charge, giving it a non-volatile memory property.

This non-volatile memory property of a memristor is the inspiration and principle behind memristive RF switches (Chua 2011). Memristive concepts were initially adopted for non-volatile data storage memory applications and some of the notable members of this family are given here in brief.

– ReRAM or RRAM (resistive switching random access memories): these are made up of memory cells that utilize the non-volatile change in resistance across a solid-state dielectric material for data storage memory applications (Chua 2011; Bocquet *et al.* 2014; Guy 2015; Leon 2018). CBRAM, oxide-based resistive random access memories (OxRAM) and PCM are a subset of this category.

– FeRAM (ferroelectric random access memory): this is a technology in which data are stored by utilizing the non-volatile dielectric polarization orientations in ferroelectric materials (Chanthbouala *et al.* 2012).

– MRAM (magnetic random access memory or magneto resistive random access memory): here, the magnetization in a stack of sandwiched magnetic and dielectric materials is utilized for non-volatile data storage (Daughton and Huang 1988).

Among these memory technologies, the concepts of CBRAM and PCM were later adopted and adapted for RF switching applications, and are explained in detail in the following sections.

1.3.3.1. *Phase change material-based RF switches*

Phase change material-based RF switches are a new upcoming innovation, inspired by the PCM applications. The center of this technology are the so-called phase change materials (Wuttig and Yamada 2007). Phase change materials are resistive materials that show a significant change in resistance upon a transition from crystalline to amorphous states and back, actuated by the application of heat pulses in a special manner. This phase transition is retained by the material even after the removal of heat pulse, thus making it a non-volatile switching technology. In this section, we discuss only the aspects of this technology related to RF switching applications to maintain brevity.

Phase change materials are integrated into the signal path of an RF circuitry to obtain an RF switch. Many material combinations have been reported to exhibit phase change characteristics, and are found suitable for RF switch applications. Some of the commonly used materials include germanium telluride (GeTe) (Gawelda *et al.* 2011; El-Hinnawy *et al.* 2014; Wang *et al.* 2014), vanadium dioxide (VO_2) (Dragoman *et al.* 2006; Hillman *et al.* 2014; Liu *et al.* 2019), yttrium hydride (Dragoman *et al.* 2006), germanium antimony telluride (GeSbTe), antimony telluride (SbTe) (Moon *et al.* 2018) and so on.

As mentioned above, the application of a controlled heat pulse results in a reversible transition of phase change material from an insulator to an electrical conductor. The magnitude, rise time and cooling time of the heat pulse determine the phase transition of these materials.

For example, phase change in the case of GeTe (El-Hinnawy *et al.* 2014) is achieved as follows, and is illustrated in Figure 1.8. Here, in a given phase of GeTe, the application of a short (~100 ns) high-temperature heat pulse, to raise the temperature above melting point, would result in liquidation of the material. In this liquid state, atoms are randomly distributed among themselves without any structural arrangements. A rapid (~100 ns) cooling at this state would freeze the material, leading to an amorphous state with very high electrical resistance, retaining the randomness of atomic distribution.

Inversely, the application of a broad pulse with a preset moderate amplitude to raise the temperature of this amorphous material above its crystallization temperature forces the atomic bonds to rearrange into a crystalline state with low electrical resistance, as illustrated in Figure 1.8. These narrow and broad heating pulses, used for phase transition, are called "quenching" and "annealing" pulses respectively.

PCM-based RF switches are developed by integrating such phase change materials into the RF signal path in series or shunt configurations, and are operated using heat pulses actuated using different electrical mechanisms. The requirements and challenges for an RF switch based on phase change materials are not different from the points stated in section 1.2. Over the past decade, a number of patents and remarkable articles have reported reliable

PCM RF switches that operate up to 100 GHz (El-Hinnawy *et al.* 2014; Hillman *et al.* 2014; Raieszadeh *et al.* 2016; Borodulin *et al.* 2017; Leon 2018). One important aspect in the design of a phase change material-based RF switch, after the selection of suitable material, is the technique for application of operating heat pulses. To the present date, different techniques are reported to be used for application of heat pulses to phase change materials as explained in the following.

In the initial stage of research in this domain, introductory designs reported using an oven for the application of operating heat pulses to the phase change materials.

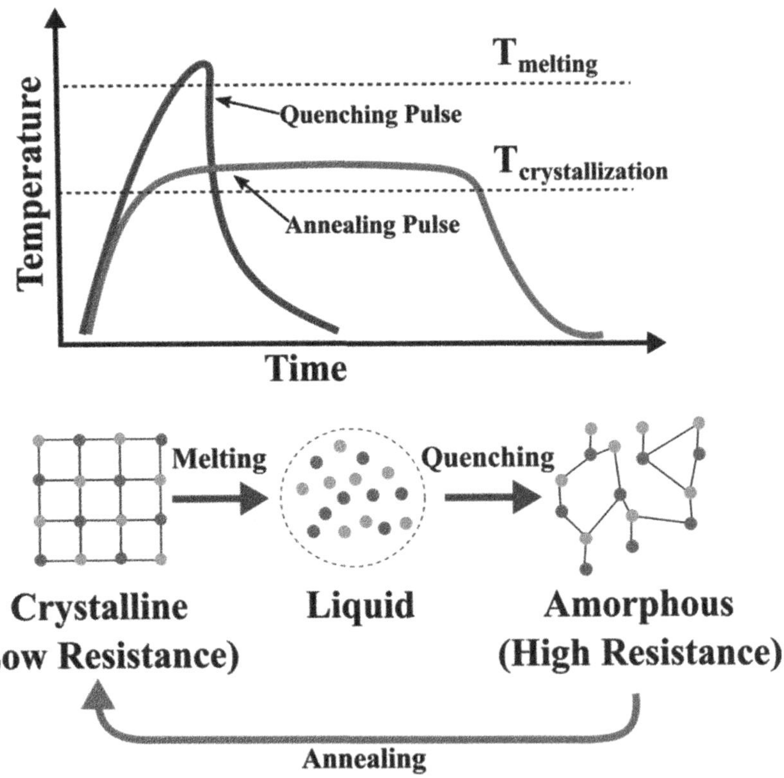

Figure 1.8. *Illustration of crystalline (low resistance) to amorphous (high resistance) reversible transition mechanism in GeTe phase change materials (idea conceived and redrawn from El-Hinnawy* et al. *(2014). For a color version of this figure, see www.iste.co.uk/jayakrishnan/technology.zip*

Nevertheless, today DC pulses for heating applications are applied to the device either directly through the RF signal path as in Shim *et al.* (2013) or using dedicated heater contacts as in El-Hinnawy *et al.* (2014). However, a design with dedicated heater contacts for application of heat pulses is simpler to implement (Mahanta *et al.* 2018). It is challenging in the former case to simultaneously maintain a low-resistance path for the RF signal and a high-resistance path for electrical resistive heating. Hence, most of the recent designs use an integrated heating element with compound alloys such as NiCrSi for the application of the heating pulses (El-Hinnawy *et al.* 2014; Wang *et al.* 2014; Raieszadeh *et al.* 2016). In addition, although not very common, some articles also have reported the use of pulsed femtosecond and nanosecond lasers for heating action (Gawelda *et al.* 2011).

Figure 1.9 presents the topology of a phase change material-based RF switch with dedicated contacts for application of heater pulses, similar to the design reported in El-Hinnawy *et al.* (2014). This design shows an ON state insertion loss less than 0.3 dB, and an OFF state isolation greater than 10 dB in the entire operating bandwidth, which is DC to 40 GHz.

At present, the phase change material-based RF switch technology is in its infancy stage, and it is progressing at a rapid rate. No commercial products are reported to date utilizing this technology for RF switching. Nevertheless, the current rate of advance in research and reliability of reported products would hopefully lead to commercial use of these switches in near future.

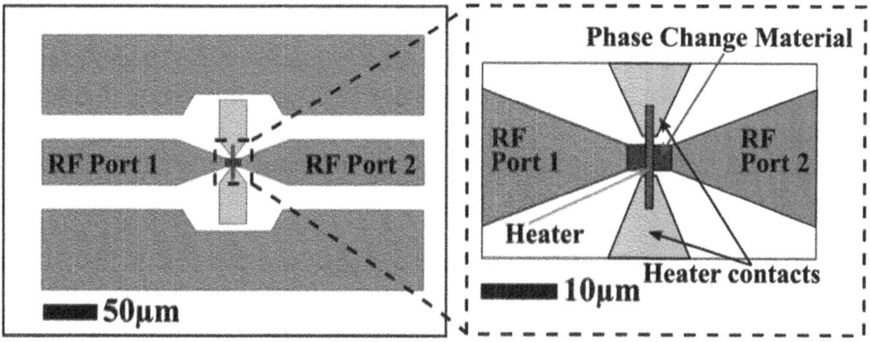

Figure 1.9. *Topology of a phase change material-based RF switch on a co-planar transmission line in series configuration. Inset shows zoomed image of switch area (idea conceived and redrawn from El-Hinnawy* et al. *(2014). For a color version of this figure, see www.iste.co.uk/jayakrishnan/technology.zip*

1.3.3.2. *CBRAM/MIM RF switches*

Application of CBRAM-based MIM switches to passive/low-power and low-cost RF microwave devices, using an optimized process compatible with mass production, is the focus topic of this book.

CBRAM/MIM switches are a subset of ReRAM. This technology depends on the memristive theory postulated by Chua (2011) and works on the principle of electrochemical formation and dissolution of a conductive filament, among two electrochemically asymmetric electrodes sandwiching a layer of solid ion conductor compound (Guy 2015).

Even though bistable resistive switching in metal–insulator–metal structures was observed implicitly or explicitly since the 1980s (Pagnia and Sotnik 1988), CBRAM technology in its current form was initially identified and invented in 1996 by Professor Michael Kozicki of Arizona State University. Professor Kozicki called these switches programmable metallization cells (PMC) and patented this technology with several variations and new innovations (Kozicki and West 1998, 2002; Kozicki 2002, 2010; Kozicki and Balakrishnan 2008). Initially, CBRAM technology was introduced as a fast and reliable non-volatile memory technology. Today several commercial memories are available utilizing CBRAM technology (Adesto Technologies 2016), and CBRAM is a registered trademark of Adesto technologies (Adesto Technologies 2020).

The relevance of this technology is well accepted even by the Defense Advanced Research Projects Agency (DARPA) of the United States Department of Defense, who acknowledged it by providing a grant to Adesto Technologies in 2010 for further advanced research on this topic (Adesto Technologies 2010). CBRAM is recognized and considered as a potential candidate to dominate future non-volatile memory applications in the International Technology Roadmap for Semiconductors (ITRS) report of 2011 (ITRS Reports 2011).

The potential of this technology as a non-volatile RF switching solution was soon revealed through some innovative research articles and patents in the recent decade (Nessel *et al.* 2008; Nessel and Lee 2011; Vena *et al.* 2012a; Pi *et al.* 2015, 2016; Xia *et al.* 2015; Dragoman *et al.* 2017). In this section, we discuss the RF and microwave switching applications of this technology. Device topology examples and state of the art in CBRAM/MIM RF switching technology are discussed separately later in this chapter.

A CBRAM-based MIM cell, as its name indicates, is formed of two electrochemically asymmetric electrodes sandwiching a layer of ion-conductor compound, which is also an electronic insulator. Figure 1.10 shows the layer architecture and working mechanism of a typical CBRAM/MIM switching cell. Here, one of the electrodes is an electrochemically active ion-donor such as silver (Ag) or copper (Cu), and the other is a relatively electrochemically inert ion acceptor such as aluminum (Al) or gold (Au). The switch action based on electrochemical formation and dissolution of a conductive filament in CBRAM/MIM cell is actuated and controlled using applied DC pulses.

The working mechanism of a CBRAM/MIM switching cell could be explained as follows. Initially, the electrical resistance across a CBRAM cell is very high due to the presence of an electrically insulating ion-conductor layer. Application of a positive DC electric voltage (with a sufficient current limit) with respect to the active electrode initiates a redox mechanism in the cell. On the basis of this, the active electrode surface in contact with the ion conductor starts to oxidize by losing electrons and liberates metal cations, as shown in Figure 1.10(a). These ions, under the influence of electric field induced due to the DC voltage, would migrate through the ion conductor layer to get reduced, by receiving electrons, to neutral metallic atoms at the inert electrode, and this process is called the forming process.

This reduction process at the inert electrode would result in the formation of a dendrite-shaped metallic filament at the inert electrode which will eventually touch the active electrode, thereby lowering the overall resistance of the MIM cell, thus establishing the "set" (or ON) state, as shown in Figure 1.10(b).

Conversely, the application of a sufficient reverse DC voltage (and current), that is, a voltage which is positive with respect to the inert electrode, would result in the reversal of this process. Here, the filament would start to dissolve due to the same redox mechanism, but this time in the reverse direction. Oxidization takes place on the metallic filament attached to the inert electrode and reduction at the active electrode. This would result in rupture of the conductive metallic filament, increasing the resistance of the MIM cell, thus establishing the "reset" (or OFF) state, as shown in Figure 1.10(c).

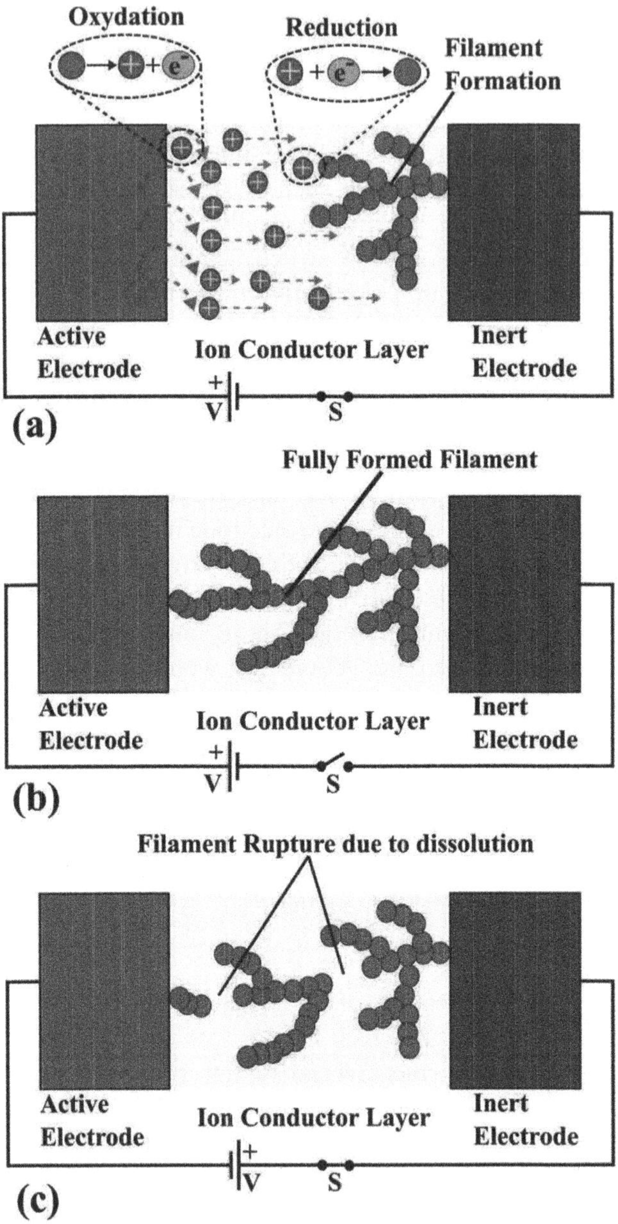

Figure 1.10. *Layer architecture of a typical CBRAM/MIM switch and its working mechanism. (a) Start of filament formation (forming process). (b) Fully formed filament (set or ON state). (c) Filament dissolution (reset or OFF state). For a color version of this figure, see www.iste.co.uk/jayakrishnan/technology.zip*

A CBRAM/MIM switch requires actuation in the form of DC pulses only to control the formation of a metallic filament and reach a desired impedance state, as explained above. The cell retains this impedance state even on withdrawal of the actuation mechanism, thus exhibiting non-volatile switching behavior. A power supply or actuation in the form of DC pulses is a requisite only when a change in present impedance state (ON or OFF) is desired.

Dissolution of metallic filament during reset process depends critically on the time duration of application of reverse voltage. A short pulse of sufficient amplitude and current (applied with right polarity) would lead to a high resistance reset state, by rupturing the filament, in its thinnest (or weakest) parts only. This is the consequence of a short pulse with high amplitude and current, which locally deforms the conductive filament at its weak parts due to localized heating. Then the actual dissolution mechanism as explained above would start. Here it is not necessary that the entire filament be dissolved back into the active electrode, but only that it be ruptured to break the current path, as seen in Figure 1.10(c). This is advantageous for the next set cycle as a rapid filament formation can be achieved by just making connections to the ruptured areas. This phenomenon brings significant difference in the characteristics of set pulses required for the first forming process and other set cycles, after the first reset following the forming process. Thus, for the subsequent set processes, a set voltage of smaller magnitude and shorter time period could be used to fill this small gap in the filament. In such a case, the gap between the inert electrode (which can be a part of a former filament connected to the inert electrode) and active electrode (which can also be a part of a former filament connected to the active electrode) is smaller than the initial gap.

One should also note that a prolonged application of reverse voltage would lead to more and more dissolution of the filament due to prolonged redox reactions (Yang *et al.* 2012). Similarly in the forming process, a prolonged exposure to set/ON voltage would lead to stronger filaments (Pi *et al.* 2015). Extensive research has been carried out to describe and observe the working mechanism of CBRAM/MIM cells, confirming the validity of above description (Nessel *et al.* 2008; Lin *et al.* 2011; Yang *et al.* 2012, 2014; Busby *et al.* 2015; Guy 2015; Pi *et al.* 2015; Yalon *et al.* 2015; Song *et al.* 2017).

It is worth noting however that, in some cases, especially for some ion conductors with low ion mobility/conductivity values (e.g. amorphous silicon [a–Si]), the direction of filament formation is reversed, i.e. the filament growth during forming and set process happens from the active electrode to the inert electrode, while all other mechanisms remain the same (Yang *et al.* 2012). Anyhow, such materials are seldom used and desired for RF switching, due to the requirement of fast switching speed and low ON state resistances.

Several metal combinations are found to be suitable as electrodes for CBRAM/MIM switching applications. This includes electrochemically active metals such as silver, copper and so on and relatively inert metals such as aluminum, gold, platinum, tungsten and so on (Pagnia and Sotnik 1988). The choice of ion conductor is a critical factor in the architecture of CBRAM/MIM switches. The choice of material and underlying process for the formation of this ion conductor layer influence factors such as switching speed, switching voltage and current, stability, cost of manufacture and so on (Yang *et al.* 2012). Several materials including doped chalcogenide glass (Kozicki and West 1998; Nessel and Lee 2011), semiconductor oxides (Vena *et al.* 2012a; Yang *et al.* 2012), ion conducting and metal doped polymers (Perret *et al.* 2015; Jian *et al.* 2017; Song *et al.* 2017), and several other material combinations (Pagnia and Sotnik 1988), including very precisely controlled air gaps (Pi *et al.* 2015), have shown ion-conducting properties suitable for CBRAM/MIM applications.

1.4. State of the art of CBRAM/MIM RF switching technology

CBRAM technology, as mentioned above, was originally introduced as a non-volatile memory technology. In addition, this technology is being investigated as an efficient and emerging RF switching solution and has produced a handful of promising results. In this section, before going into the details of state of the art accomplishments in this filed, it is interesting to mark the important aspects of expected performance parameters distinguishing among different applications of CBRAM technology: memory and RF switch. Table 1.2 summarizes some of the significant unique features (and their necessities) demanded by each of these applications of CBRAM/MIM technology.

CBRAM technology – memory application	**CBRAM technology – RF switch application**
Typically nanometric-sized cells.	Typically micrometer-sized cells.
ON and OFF states require only a notable distinction. A few kΩ at ON and a few MΩ at OFF states are well acceptable.	ON states should be necessarily less than a few ohms (e.g. <10 Ω) and OFF states should be greater than a few kΩ to ensure proper RF transmission and isolation.
Ion-conductor thickness is typically a few nanometers.	Ion-conductor thickness varies typically from a few nanometers to a few micrometer depending on application.
Write and read speeds are critical.	Linear behavior, RF compatibility and switching speed are critical.
Power handling capability is not a crucial factor.	Power handling capability is a crucial factor.
Packing density on a single wafer in a given space is vital.	Reliability as a stand-alone switching device is vital.

Table 1.2. *Performance parameter features distinguishing requirements for memory and RF switch applications of CBRAM/MIM technology*

RF switches utilizing CBRAM/MIM technology and operating up to 40 GHz have been recently proposed in the literature. Notable contributions identified as the state of the art include nanoionic RF switches based on silver (Ag)-doped germanium selenide ($GeSe_2$), invented by James Nessel *et al.* of NASA Glenn Research Center, USA (Nessel *et al.* 2008; Nessel and Lee 2011), and nanoscale memristive RF switches based on a 35 nm air gap invented by Shuang Pi *et al.* of the University of Massachusetts, USA (Pi *et al.* 2015, 2016; Xia *et al.* 2015). These two articles share the same principle of operation, but are unique in the choice of selection of ion-conductor for switching applications. The former uses a chalcogenide compound and the later use a precisely controlled air gap as the ion-conductor.

The work of Nessel *et al.* proposes a planar CBRAM/MIM switch topology on an RF transmission line, as shown in Figure 1.11 (recreated from Nessel *et al.* 2008).

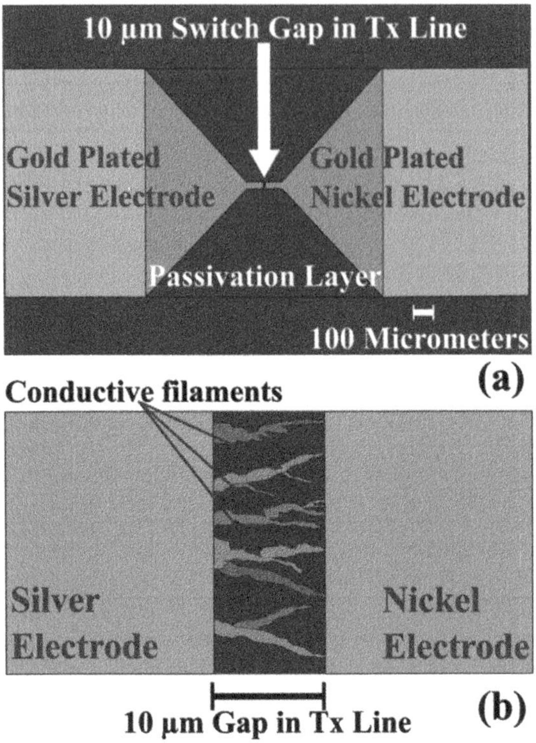

Figure 1.11. *Nanoionics-based RF switch (redrawn from Nessel et al. 2008). (a) Topology of nanoionic RF switch and (b) illustration of microphotograph showing filament formation among active (silver) and inert (nickel electrodes). For a color version of this figure, see www.iste.co.uk/jayakrishnan/technology.zip*

Here, a 10 µm wide planar gap in a RF transmission line filled with a chalcogenide composite of silver (Ag)-doped $GeSe_2$ as the ion conductor functions as the CBRAM/MIM switching mechanism. Gold-plated silver (Ag) and nickel (Ni), respectively, form the active and inert electrodes of the CBRAM/MIM cell in this device, as shown in Figure 1.11. Here, the CBRAM switch in summary operates in series configuration on the parent transmission line.

DC pulses of amplitude 1 V (ON process) and –1 V (OFF process), both with a compliance current limit of 10 mA, are used to operate the device. DC pulses are fed directly through the RF signal path using appropriate isolation networks. The device reports an ON or set state impedance around 10 Ω, and a higher resistance at OFF or reset state sufficient to provide a good RF isolation.

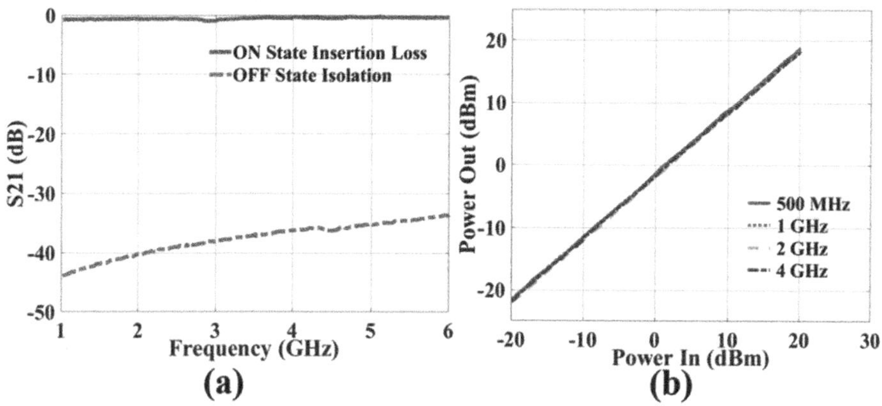

Figure 1.12. *RF performance characteristics of nanoionics-based RF switch shown in Figure 1.11 (redrawn from Nessel et al. 2008). (a) RF ON and OFF state insertion loss and isolation. (b) Relation among input and output RF power showing linearity characteristics. For a color version of this figure, see www.iste.co.uk/jayakrishnan/technology.zip*

Illustrations of microscope images exhibiting silver resistive filament in the ion conductor layer, formed as a result of switching action, among the silver active and nickel inert electrodes are shown in Figure 1.11(b) (redrawn from Nessel *et al.* 2008), validating the theory of CBRAM/MIM switching cells. This switch shows an operational frequency range of 1–6 GHz, with ON state insertion loss better than 0.5 dB and OFF state isolation better than 30 dB in the entire bandwidth as shown in Figure 1.12(a) (redrawn from Nessel *et al.* 2008). The switch also shows a linear operation in the given bandwidth for RF input power levels from –20 dBm to +20 dBm, as shown in Figure 1.12(b) (recreated from Nessel *et al.* 2008). Authors claim a switching speed of 1–10 μs for this device. Details on number repeatability of switching cycles or endurance cycles of the presented device are unknown.

The next state of the art design in this field is the work of Shuang Pi *et al.* This is a nanoscale memristive RF switch composed of gold (Au) and silver (Ag) electrode pairs separated by a 35 nm air gap, as shown in Figure 1.13 (redrawn from Pi *et al.* 2015). Low capacitance due to a lower dielectric constant of the air gap, in combination with chosen design geometry, provides an operating range of up to 40 GHz for this device.

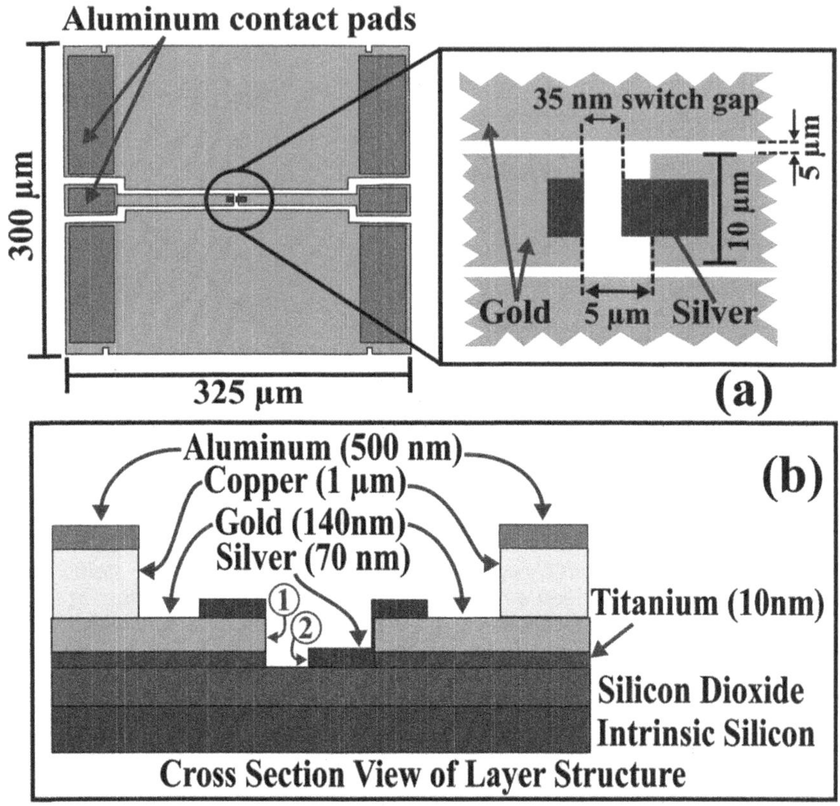

Figure 1.13. *Nanoscale memristive RF switch (redrawn from Pi* et al. *2015). (a) Topology of memristive RF switch, with zoomed inset showing 35 nm CBRAM/MIM switch gap. (b) Cross-section view showing layer structure of memristive RF switch. The filament formation is between gold and silver electrodes and marked as ① and ②. (Please note that the image is only an illustrative representation and the layers are not up to scale. Please refer to legends for dimensions). For a color version of this figure, see www.iste.co.uk/jayakrishnan/technology.zip*

Authors claim an ON state insertion loss less than 0.3 dB and an isolation greater than 30 dB from 10 to 40 GHz, respectively, for ON and OFF states of the switch, as shown in Figures 1.14(a) and (b) (replotted from Pi *et al.* 2015). DC voltage pulses with the mentioned polarity corresponding to the silver (active) and gold (inert) electrodes, respectively, are used to operate the device. Non-returning sweeps in the range of 0 to +4 V are used for the ON process and in the range of 0 to –0.4 V for the OFF process. A maximum compliance current limit of 50 mA is imposed for these voltages.

An average ON or set state resistance value of 3.6 Ω and a high resistance state in OFF or reset state which provides a good RF isolation is reported for this realization. This device also reports a linear power handling capability up to +17 dBm for the switch, above which the RF power is seen to induce self-switching characteristics. The maximum operating frequency of this switch is computed through full-wave electromagnetic simulation and reported to be 200 GHz, exhibiting ON state insertion loss and OFF state isolation values of less than 0.6 dB and greater than 17 dB, respectively, for the entire bandwidth. The presented device shows a switching cycle repeatability of around 80 cycles in a natural environment (which could be extended to around 700 cycles by protecting the switch area in deionized water) and a theoretically calculated ON/OFF state retention time of 3.3 years, as reported by the respective authors.

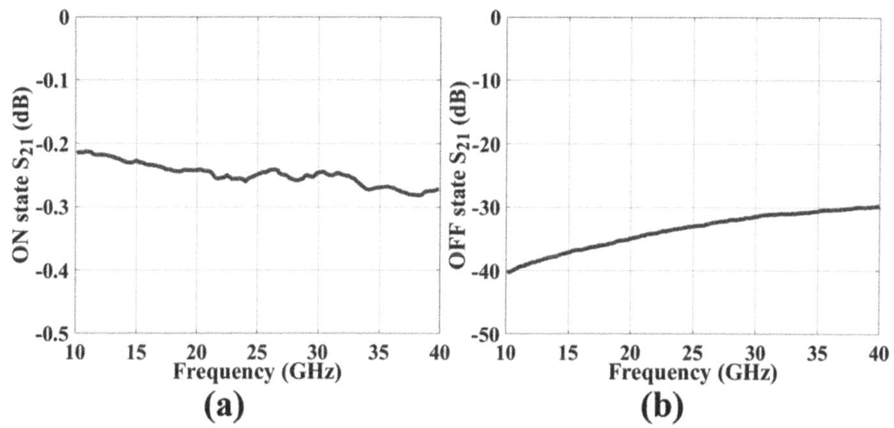

Figure 1.14. *RF performance characteristic of memristive RF switch, shown in Figure 1.13 (redrawn from Pi* et al. *2015). (a) ON state insertion loss. (b) OFF state isolation*

The presented state of the art devices present the capability and achievable performance of CBRAM/MIM technology for RF switching, even in its initial stage. Like in the infancy stage of any technology, CBRAM/MIM technology also has some challenges to be revamped through which these devices could promise a stunning future for RF switching.

One could note here that fabrications of both prototypes discussed above are accomplished using clean room procedures, and the available information on switching cycle repeatability or endurance shows that improvement is required in this aspect. Some of the improvements we try to address in this book include the introduction of simple processes for fabrication of non-volatile CBRAM RF switches and passive or low power RF and microwave devices with integrated CBRAM switches, through techniques compatible with mass production, at low cost. These processes should utilize materials that could be handled and operated in ambient environments, and not only in clean rooms, and free from any high temperature curing. Adaptation of these devices for an improved number of switching cycles is also a foreseen improvement.

1.5. Demand for a non-volatile RF switch and selection of CBRAM/MIM technology

In the previous sections, we have discussed different types of RF and microwave switching technologies their merits and demerits. From this discussion, it is clear that the choice of a switching technology is a compromise among different parameters such as type of application, required performance (e.g. switching speed), power consumption, operating bandwidth, cost and so on.

In this section, we discuss the potential areas of demand and requirement for a low-cost and non-volatile RF switching solution. An increase in demand for a non-volatile and fully passive low-cost RF and microwave switch has been intensified at present. Recent advances in the Internet of Things (IoT) backed by fifth-generation cellular communication (5G) technology have added to this yearning for a new breakthrough (Borkar and Pande 2016; Collela 2017).

The IoT is a technology that provides Internet connectivity to physical objects in our everyday life (Attaran and Rashidzadeh 2016; Borkar and Pande 2016). The 5G cellular network standard is a powerful broadband communication technology, a notable contribution of this era. 5G could be the foundation for implementation of the full potential of IoT. This combination would be a unique solution with high-speed connectivity, very low latency and ubiquitous coverage (Collela 2017). This would support smart homes, vehicles, transport infrastructure and so on, ranging from connected home appliances like a smart air conditioner at home, smart medical care like a smart band aid communicating the current condition of a wound, smart vehicular transport systems such as trucks and buses moving without human drivers on a busy highway, and even remote control of heavy machinery in hazardous and unreachable environments. These developments would contribute to improving quality of life, speed of service and safety standards. This technology is also very near to efficient extension of this idea to remote surgery and remote monitoring of implanted medical devices. All this is expected to be done through a group of smart wireless sensors, data processors and communication links working in harmony simultaneously.

It is estimated that more than 50 billion devices will be connected to the Internet in 2020 worldwide (Broadband Wireless Networking Lab 2019). However, for all these wireless systems to be truly "wireless" all the time, they should be passive or self-sufficient in terms of operating power supply. To meet such power supply or power storage regulations, one of the requirements is that these systems require a passive or low-power switching solution for needs such as reconfigurable antennas (which include 3D steerable MIMO systems, frequency agile antennas and similar), filters, resonators, object identifiers, stand-alone sensors, channel switches, power splitters and so on.

5G technology demands different types of RF switches, for example low-power consumption and high-performance switches for mobile phone base stations, low-cost non-volatile switches for battery-less or low-power stand-alone sensors and passive transponders and so on. At present, the former requirement (low-power consumption and high-performance switches) could be fulfilled using standard commercial CMOS switching devices. However, the latter requirement, i.e. low-cost non-volatile switches, is our area of interest of application of CBRAM/MIM switches.

A fundamental difference from 4G and one of the capital expectations for 5G communication technology is to provide connectivity to "things" through IoT concepts, utilizing smart low-power/passive sensors and transponders, attached to a connectivity seeking object. The object could in reality be anything which would like to pursue a certain degree of "smartness", such as an electronic home appliance requiring remote control, an everyday object like a wallet that could be tracked, a food package that could monitor the quality of its contents or a medical wound dressing that provides information on the status of a wound.

A very simple and primary step in making any object "smart" or connected is about attaching a "smart-label" that can identify itself, sense certain parameters and communicate to a base station. To keep it simple, a smart packet of yogurt or ice-cream could, upon request, identify itself to your mobile phone when proximal, tell you where it is in a supermarket, tell you its flavor, price, about its condition such as if it is consumable or rotten, etc.

Recent trends and advances in research in the field of passive RF and microwave devices show that RFID systems and especially chipless RFID systems could serve the purpose of requirements such as object identifiers and stand-alone sensors (Vena *et al.* 2011; Nair *et al.* 2013; Perret *et al.* 2013; Tedjini *et al.* 2013; Amin *et al.* 2014a; Perret 2014; Feng *et al.* 2015; Karmaker 2016; Bibi *et al.* 2017; Borgese *et al.* 2017; Vena *et al.* 2016). Chipless RFID technology could be used as a smart label which does not require any power supply or battery. It is often called the RF-Barcode due to similarities in appearance and operation to an optical barcode system, however in contrast it has a lot of additional functionalities such as non-optical line of sight reading (readability even if it is visibly hidden), electronic rewriteablity, sensing capabilities and so on.

Electronic rewriteablity of a chipless RFID tag is a unique feature which we try to address through this book. Electronic rewriteablity provides enormous flexibility and exceptional dominance to chipless RFID technology as a smart label which appreciates tag reuse, enabling mass production of similar blank tags in large quantities through techniques such as roll-to-roll processes, which could be pre-coded or written/rewritten to encode a desired identifier at time of application. In Chapter 3, we explain innovative approaches to the development of electronically rewritable chipless RFID tags with integrated CBRAM/MIM switches, with processes compatible with mass productions.

In addition, a lot of innovations have also been proposed recently in the field of pattern steerable and frequency agile antennas including MIMO systems, and in electronically reconfigurable filters (Zhang and Zhu 2011; Chaudhary *et al.* 2012; Ding and Wang 2013; Dragoman *et al.* 2017; Lin *et al.* 2017; Chen *et al.* 2018; Al-Yasir *et al.* 2019). These research advances are also backed by multidisciplinary engineering technologies including smart materials (Amin *et al.* 2014b; Bibi *et al.* 2016), 3D printing and conductive inkjet and thermal printing techniques (Cook *et al.* 2013; Ellinger *et al.* 2015; Chan *et al.* 2018; Ellinger *et al.* 2018; Saracho-Pantoja *et al.* 2018).

On the other hand, most of these devices cited above utilize classic volatile switching technologies such as solid-state semiconductors (PIN diode switches, MEMS switches, hybrid switches and similar) for electronic reconfigurability or mechanical solutions such as laser ablation and localized inkjet printing and similar in reconfigurable chipless RFID applications. The former (semiconductor switches) add to the power budget of the system, and furthermore make the presence of an onboard battery pack mandatory. The latter (mechanical solutions) are not efficient and reliable techniques.

For fulfilling the switching requirements of these devices, apart from commercially available volatile switching solutions such as solid-state semiconductor or MEMS switches, these switches should have certain unique characteristics as follows: non-volatile retention of impedance state, low power consumption, long time switch impedance state retention (stability), easy integration into target device and so on. A non-volatile RF switch with these desired characteristics would benefit every field of RF and microwave engineering, especially the above-stated domains, as illustrated in the form of a chart in Figure 1.15.

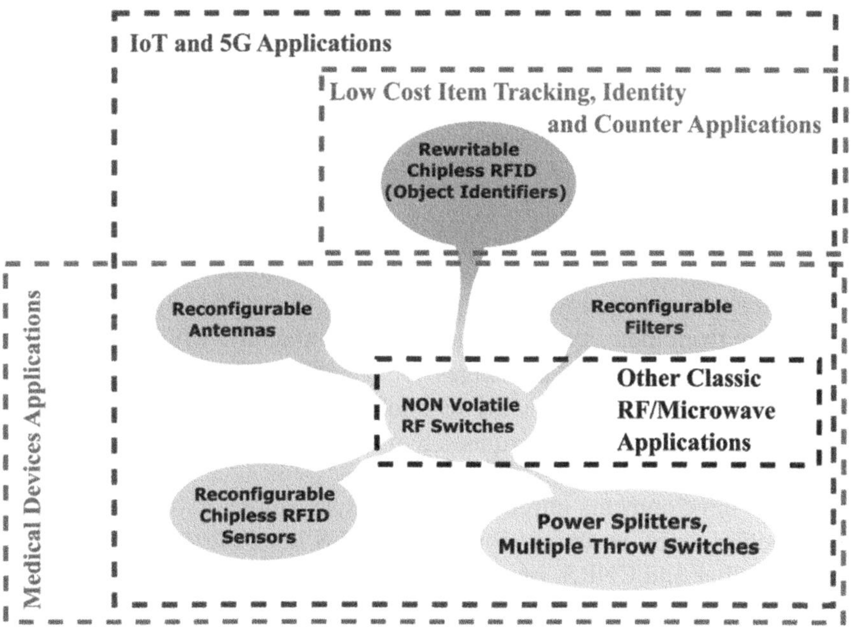

Figure 1.15. *Core areas of application of a non-volatile RF switch. For a color version of this figure, see www.iste.co.uk/jayakrishnan/technology.zip*

In this book, importance is given to investigation and optimization of a low-cost non-volatile RF switch, which could be directly integrated into an RF/microwave device or circuit, using simple processes compatible for mass production. Such an approach has never been produced up to now (to the best of our knowledge), despite the research for revealing best performance and potential theories related to any particular switching technology. Here we ask ourselves and try to answer the following questions: is it possible to print a switch with a simple off-the-shelf printer? Or at least is it possible to realize a low-cost non-volatile RF switch (integrated into RF devices) using simple and rapid industrial technology used for printed electronics? Both classic electronic substrates and flexible substrates like PET are expanded as Polyethylene terephthalate or paper to call these devices the smart labels of future.

To answer these challenges, we choose the CBRAM/MIM switch technology for the following reasons. Out of the two main available solutions for non-volatile RF switching technology, PCM and CBRAM/MIM, it appeared to us that CBRAM/MIM is a simpler technology, compatible with mass production in an industrial setup. The proof of concept of this simplicity of CBRAM/MIM technology is our primary goal through this research. Thus in contrary to previous studies in this domain, we try to show that simple materials (that exist not only in clean room technologies), through uncomplicated processes, could be utilized for establishing a low-cost switching solution based on CBRAM/MIM technology.

As of now, PCM technologies necessarily require clean room microfabrication technologies, high temperature processes and even joule heating simulations setups, which span a great deal of tedious requirements all the way from their design procedure to manufacture, which imply a great spectrum of multidomain challenges for practical industrial processes. It is worth comparing CBRAM/MIM technology to PCM at this point, which could meet these challenges exquisitely in a much adroit approach.

Recent developments introduced at the LCIS laboratory of the University of Grenoble Alpes, in association with the IES laboratory of the University of Montpellier and CNRS, on CBRAM/MIM switch technology, bring hope and a new horizon for low-cost and passive reconfigurable electronics compatible with mass production in industrial setups (Vena *et al.* 2012a;

Perret *et al.* 2015; Jian *et al.* 2017) and for an inventive development of electronically reconfigurable passive RF devices.

Now, to continue the comparison of CBRAM/MIM and PCM, an Achilles heel in the design of PCM switches is the requirement of a heater for the phase change process of the switching material from amorphous to crystalline and back. About the complexity of the realization technique, we can note that the control part requires the use of transducers to apply the heating pulse, which greatly complicates the realization. This heater path is different from the RF signal path in the case of indirect heating. In such indirect heating, a DC electrical signal heats an intermediate highly resistive material that transfers its calorific energy to PCM via the Joule effect, which requires a complicated multilayer structure. Conversely, we can consider the techniques of direct heating, in which the heat pulses are applied directly through the RF path. A smart material for the heating process must be found that could simultaneously maintain a low resistance path for the RF signal and a high resistance path for resistive heating, which again complicates the system (Mahanta *et al.* 2018).

However, it is true that after a lot of work done by different teams from all around the world, PCM technology is now more mature, and is still advancing at a good rate. Still, from the authors' point of view, the weightage of simplicity in fabrication is more on the side of CBRAM/MIM technology due to the following reasons.

PCM switches are based on the use of some very specific materials (GeTe for example). This type of material requires a series of sophisticated process steps for layer formation; and also needs a passivation layer to protect it from the environment; if not, the material is not stable in an ambient atmosphere. Up to now, it seems very difficult (even impossible) to print this material. At the same time, several studies (Vena *et al.* 2012a; Perret *et al.* 2015; Jian *et al.* 2017) indicate a ray of hope for printing a CBRAM/MIM switch layer by layer using an inkjet printer, utilizing metallic inks, and polymer ion-conductor solutions, such as poly methyl methacrylate (PMMA).

Such a breakthrough would be a great advantage for printed passive stand-alone RF and microwave devices such as rewritable chipless RFID-based identifiers and sensors. This would also be a great step in the integration of these switches without any techniques like soldering, in directly printed RF and microwave circuits, realized with the help of inkjet or 3D printing technologies (Cook *et al.* 2013), even on low-cost and flexible substrates like paper (Vena *et al.* 2013a; Borgese *et al.* 2017; Vena *et al.* 2018) or PET (Ali *et al.* 2019).

Such a non-volatile RF switch could find applications in the following domains:

– providing electronic rewriteablity to chipless RFID tags for identification and sensing, and manufacture of such tags (smart labels) on flexible substrates like paper;

– providing electronic reconfigurability to filters and antennas, which include systems such as frequency reconfigurable and tunable filters, pattern steerable antenna systems, frequency agile smart antennas and so on. Mechanically complicated designs like reflect and transmit arrays and MIMO systems could be printed directly along with integrated switches on flexible substrates, which could then be reconfigured electronically, and mechanically oriented, utilizing the flexible nature of substrates;

– realizations of electronically reconfigurable multiple throw switches and power splitters;

– low-cost smart wireless sensors: these devices are an intelligent combination of all three above-mentioned applications, and would find great demand in IoT applications. Such passive transponders handling sensing and communication require electronically reconfigurable and non-volatile switches for antenna, filter and identifier (chipless RFID) requirements.

In this book, we present a summary of our efforts and affirmative outcomes of an investigation of the possibilities of design and fabrication of integrated non-volatile CBRAM/MIM RF switches for passive and low-power applications through simple processes compatible with mass production in an industrial setup on classic and flexible substrates. Table 1.3 summarizes desired improvements addressed in this book in switching performance and fabrication technologies for CBRAM-based RF switches in comparison to limitations faced by reported state of the art devices in this field.

Limitations faced by present state of the art technology for CBRAM RF switches	Targeted and desired improvements for CBRAM RF switching technology
Low switching cycle repeatability achieved (<1,000 cycles).	Improved switching repeatability suitable for target application.
Fabrication techniques are mainly "clean room"-based processes.	CBRAM RF switches compatible with low cost industrial mass productions in ambient environment.
Inorganic materials that require sophisticated high temperature processes are used as ion-conductors.	Fabrication of CBRAM RF switches using (easily) moldable polymer-based ion conductors that do not require a high-temperature process.
No commercial products (RF switches) to date.	Low-cost techniques and materials compatible with mass commercial production.

Table 1.3. *Targeted and desired improvements for CBRAM RF switching technology addressed in this book*

We believe that we have answered some parts of the above-mentioned issues successfully through the work and outcomes presented here: is it possible to print a switch with a simple off-the-shelf printer, or at least to realize a low-cost non-volatile RF switch using simple and rapid industrial technology used for printed electronics on rigid and flexible substrates like paper?

The answer here is yes! We have affirmatively proved the feasibility of realizing low-cost non-volatile RF switches and passive RF devices integrated with these non-volatile switches. The CBRAM/MIM switching technology is utilized in simple and rapid process steps compatible with mass production, on rigid as well as flexible substrates like paper. The presented switch technology and chosen materials are potentially printable, and the optimization for this is progressing quickly.

1.6. Conclusion

In this chapter, we discussed the motivation for and background to the need for a low-cost non-volatile RF switch technology. We presented in brief an idea about the requirements and expected performance factors of a switch at RF and microwave frequencies. Then we gave an overview of the concepts and advances of prominent RF switching technologies available at present. The state of the art in CBRAM/MIM switch technology was also presented. After that, we discussed in detail the requirement of a low-cost and non-volatile RF switch and reasons for the choice of CBRAM/MIM switch technology for applications addressed in this book.

We have also discussed in detail the versatile applications of CBRAM/MIM RF switches, using the proposed fabrication processes in this book, in the domain of IoT and 5G applications. CBRAM/MIM switching technology is a simple approach for a new generation of low-cost RF switching solutions. The simplicity in realization and potential printability of this technology provides a good contrast from other similar counterparts such as PCM. Physical functioning of CBRAM/MIM switches are straightforward to understand and easy to analyze, and so is the control over electrical parameters such as ON/OFF state impedance and OFF state capacitances, which play a critical role in RF performance.

The layer architecture of CBRAM/MIM cells makes it easy to integrate it among a standard RF circuit to achieve electronic reconfigurability, without much perturbation to the natural working of the circuit. Moreover, the optimization of such devices could be done directly in electromagnetic simulation platforms, without the direct requirements of multiphysical parameter simulation support.

The ion-conducting performance observed in moldable polymers like PMMA is a crucial factor in simplifying the realization of RF devices with integrated CBRAM/MIM switches, as explained in this book. Here, we introduce the use of a much more efficient and innovative ion-conducting polymer called nafion for CBRAM-based RF switching applications. The most endorsed specialty of these polymer ion-conductors over classic compounds like chalcogenides is their availability in form of resin solutions. This helps in growth/formation of layers of desired thickness through simple processes like spin coating. These polymers do not require any high-temperature processes for curing and are stable in ambient environments. In the following chapters, we present the integration of CBRAM/MIM switches, using such polymers (Nafion), into passive microwave devices such as chipless RFID tags, antennas, filters and SPDT switches, making them non-volatile electronically reconfigurable devices. The reasons for the choice of nafion and its advantages are explained in detail in Chapter 2.

With the confidence gained from the outcomes of experimental results obtained in this study, and the considerable RF performance quality of presented devices, we optimistically predict the development of fully printed electronically reconfigurable passive RF devices, realized using common off-the-shelf printers, commercially as well as for domestic use, in the near future. This would be done on flexible substrates such as paper, utilizing metallic and ion conductor inks.

In the next chapter, we present the real-world implementation challenges of a low-cost non-volatile RF switch, and the proposed techniques to efficiently overcome them. In Chapter 2 we propose simple realizations of CBRAM/MIM RF switches that could be undertaken in an ambient environment using processes compatible with mass production. These are validated through the realization of CPW RF switches on rigid and flexible substrates and extension of the application to an SPDT switch.

2

Real-World Implementation Challenges of a Low-Cost Non-Volatile RF Switch

2.1. Introduction

In this chapter, we address the real-world implementation challenges of a low-cost non-volatile radio frequency (RF) switch and possible solutions for the same. A reliable and printable low-cost non-volatile RF switch would be a giant step in microwave/RF engineering in this era, this also being a great boost in development of printable passive electronic devices (Cook *et al.* 2013; Vena *et al.* 2013a; Amin *et al.* 2014a; Perret 2014; Gregory and Werner 2015; Borgese *et al.* 2017). The idea of printing a microwave device like a rewritable/reconfigurable chipless RFID tag or an antenna using a household printer or at least an industrial printer would contribute to reducing the cost and revolutionizing the technology, in comparison to the current market. Classic solutions for RF switching such as PIN diodes (Hong 2009; Chaudhary *et al.* 2013; Chen *et al.* 2018), field effect transistors (FETs), varactor diodes and micro-electromechanical systems (MEMS) (Goldsmith *et al.* 1998) have proved their mettle with smooth performance up to THz, but one factor that still limits their use in passive RF devices is the need for a continuous power supply to maintain their impedance state. Moreover, these devices in most cases require a costly "clean room" process for their fabrication, and techniques such as conductive gluing or soldering to integrate them to a device. These techniques involve multistep processes both for fabrication of switches and integrating them into desired target devices.

In this chapter, we probe the possibilities of a non-volatile switching solution for RF applications based on the conductive bridging random access memory (CBRAM) technology. We present simple realization techniques compatible with industrial processes and notably without the use of any clean room techniques. We show an application of the proposed switch in the form of a single pole double throw (SPDT) power splitter. Discussions on aspects like electrical equivalent modeling, time stability, control of impedance states and other interesting factors are also discussed in detail in this chapter.

2.1.1. *Conductive bridging random access memory switches based on nafion as ion conductor*

To quickly recap, CBRAM is a solid-state switching technology evolved from the programmable metallization cell (PMC) theory (Kozicki and West 1998) developed by Professor Michael Kozicki of Arizona State University in 1996. This technology was primarily introduced as a non-volatile memory technology, but was later identified to have a potential to be used as an RF-switching technique (Nessel and Lee 2011; Vena *et al.* 2012a; Gregory and Werner 2015; Perret *et al.* 2015; Pi *et al.* 2015; Dragoman *et al.* 2017) as explained in detail in Chapter 1. The basic topology and electrical model of a typical CBRAM/MIM switching cell is given for reference in Figure 2.1.

A conductive bridging memory cell/switch includes an MIM cell, as shown in Figure 2.1. The MIM cell consists of an active electrode such as copper or silver (which is basically an ion donor) and a relatively inert electrode such as aluminum or gold, sandwiching an electrolyte layer, which is an ion conductor and electronic insulator. The classically used ion conductors include chalcogenide glass doped with group I elements, silicon dioxide and even precisely controlled air gaps (Kozicki and West 1998; Nessel and Lee 2011; Pi *et al.* 2015; Dragoman *et al.* 2017).

The switch is set on the application of a sufficiently high electric field (~1e7 to 2e7 V/m, i.e. ~10 V for ~1000 nm) between the electrodes as shown in Figure 2.1(b). This electric field causes ions from the active electrode to migrate through the electrolyte layer to the passive electrode and to form a metallic filament establishing the set (ON) state. Similarly, an electric field of opposite polarity is used to reverse this migration and break the filament establishing the reset (OFF) state (Figure 2.1(c)).

Figure 2.2 shows the microphotograph of different stages of filament formation in a typical CBRAM cell sample. These observations are an outcome of a control experiment carried out to understand the filament formation and dissolution mechanics in a CBRAM switch. The Appendix explains this control experiment in detail. Here, we use copper as the active electrode and aluminum as inert electrode and a polymer called nafion as the ion-conductor layer. The device has a separation of 100 μm between the active and inert electrodes and the nafion polymer is in a gel state (reasons for chosen material properties are a part of the control experiment and are explained in detail in Appendix A.1). These clips are extracted from a micro-videograph captured using a transmission optical microscope using the details given in Appendix A.1. The reset function is also observed in this sample, but the obtained filament dissolution was so feeble that it could not be resolved with the available microscope. These observations are seen to validate the above-stated mechanism of set–reset process in CBRAM cells.

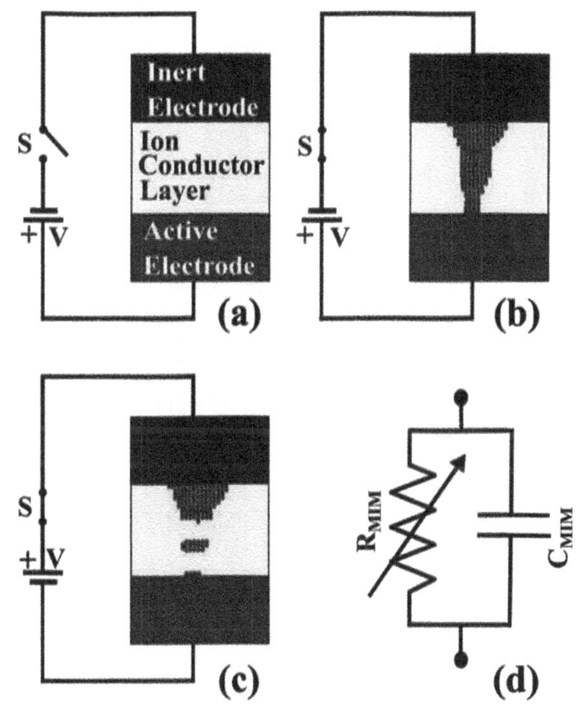

Figure 2.1. *(a) The conductive bridging CBRAM/MIM switch, (b) SET, (c) RESET and (d) electrical equivalent circuit. For a color version of this figure, see www.iste.co.uk/jayakrishnan/technology.zip*

The metallic switch link or filament formed this way is "non-volatile", i.e. the formed filament and hence the impedance state of the switch is retained even after the electric field is withdrawn. This particular feature distinguishes CBRAM/MIM switches from classically used RF switches such as the MEMS and solid-state semiconductor switches such as PIN diodes, which require a power supply to maintain their state. The electrical equivalent circuit of the MIM cell could be modeled as shown in Figure 2.1(d). Here, R_{MIM} is the equivalent DC resistance of the cell in set/reset states and C_{MIM} is the equivalent capacitance due to the geometry of the electrodes and ion-conductor used in the switch.

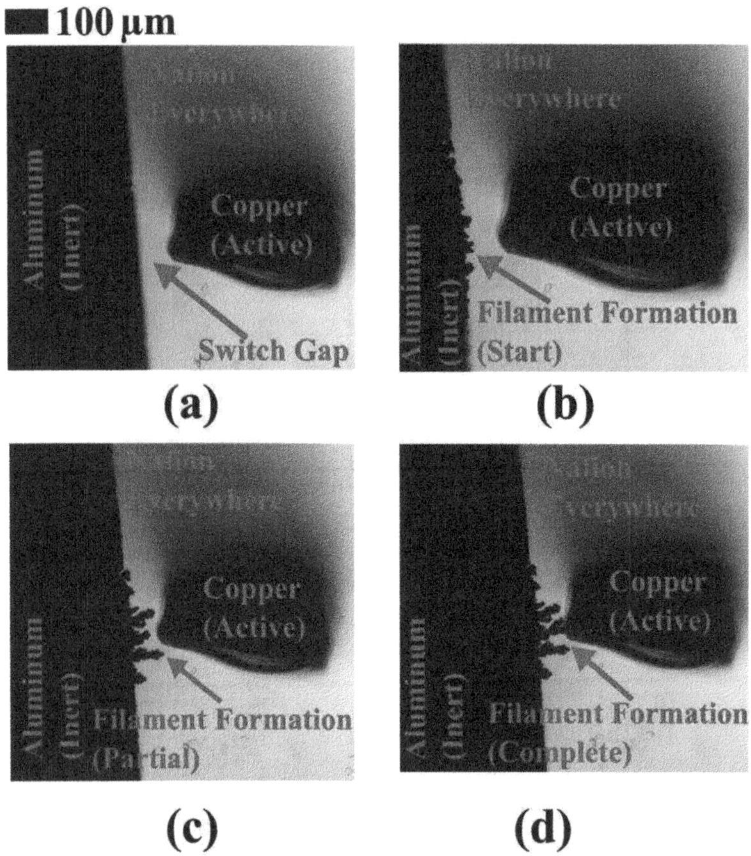

Figure 2.2. *Filament formation in a copper–nafion–aluminum planar CBRAM/MIM switch. (a) The planar MIM structure. (b) Start phase of filament formation. (c) Partially formed filament. (d) Fully formed filament. For a color version of this figure, see www.iste.co.uk/jayakrishnan/technology.zip*

CBRAM/MIM switches belong to the PMC family (Chua 2011) and are the simplest members in terms of fabrication (ease, cost and complication) and functioning (stability, fragility) from among the other members of their family, like phase change memory (PCM), which is also a proven candidate for RF switching applications (El-Hinnawy *et al.* 2014; Wang *et al.* 2014; Mahanta *et al.* 2018). In comparison to the well-established switching technologies such as the MEMS, PCM and solid-state semiconductors such as PIN diodes, the CBRAM has both some specific advantages and some shortcomings. A detailed discussion on the choice of CBRAM as an RF switch in this study has already been covered in Chapter 1.

A general comparison of the above-discussed switch technologies is given in the form of a radar diagram on a "five-point" scale in Figure 2.3, where point 5 marks the maximum advantage. It should also be noted that the conductive bridging technology for RF applications is in its budding stage and not mature enough for high-frequency and high power operations.

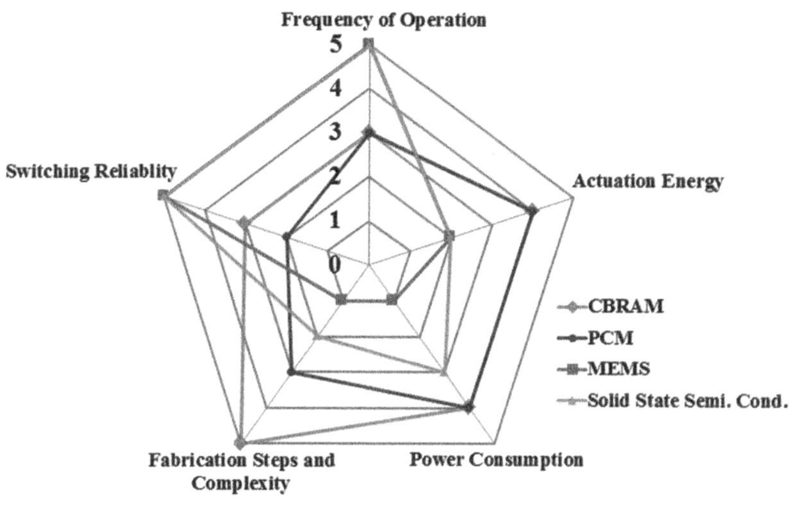

Figure 2.3. *A general comparison of CBRAM technology with other well-established RF switch technologies. For a color version of this figure, see www.iste.co.uk/jayakrishnan/technology.zip*

Among the points of comparison of PCM and CBRAM technologies for RF switching, the most interesting advantage of CBRAM technology is the ease of fabrication. A CBRAM switch could be fabricated using very simple fabrication processes on classic substrates in ambient laboratory conditions. The credit for the possibility of such simple and potentially printable fabrications and designs of CBRAM/MIM switches goes to ion-conducting polymers (Perret *et al.* 2015; Jian *et al.* 2017; Song *et al.* 2017), in place of classically used materials like chalcogenide glass (Kozicki and West 1998; Nessel and Lee 2011). This is one of the primary innovations that we propose in this book. These polymers could be transformed to stable layers of desired thickness very easily, and without any high temperature processes, as described in this section. The potential printability of these polymer layers is also a scientifically interesting feature.

From previous research studies within our group, ion conducting polymers such as PMMA and nafion are identified as good solid-state electrolyte/ion-conductors for the conductive bridging MIM cells (Perret *et al.* 2015; Susanti 2016; Jian *et al.* 2017). Among these two polymers, nafion is chosen for experiments in this work due to its superior and renowned ion-conducting properties and greater ease of handling in an ambient room.

Nafion is a registered trademark of E. I. du Pont de Nemours & Co. (Heitner-Wirguin 1996). This polymer is a sulfonated tetrafluoroethylene-based fluoropolymer copolymer that is well known for its fast-ion conducting properties and use as proton-exchange membrane in hydrogen fuel cells (Heitner-Wirguin 1996). The chemical formula of nafion is $C_7HF_{13}O_5S.C_2F_4$ and its chemical structure is depicted in Figure 2.4. Nafion has some unique advantages in terms of the ease of fabrication over inorganic compounds such as the doped chalcogenides classically used (Nessel and Lee 2011) in CBRAM memories and switching devices.

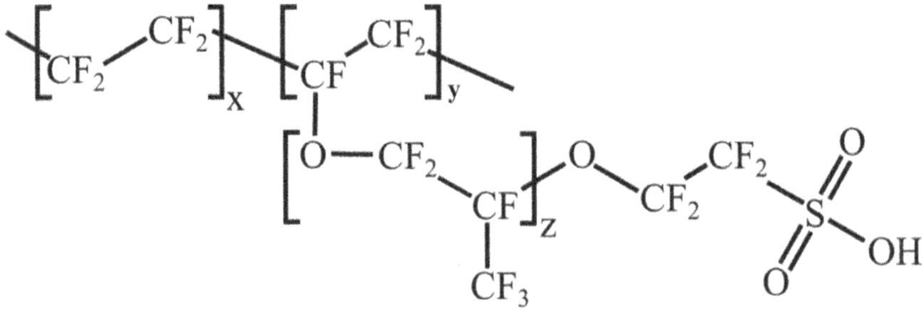

Figure 2.4. *Chemical structure of nafion*

The nafion layer could be formed by just spin coating and air drying in room temperature the nafion impregnated resin solution (Jayakrishnan *et al.* 2017, 2018a, 2018b). This would lead to saving time and cost in comparison to other classically used ion-conductors, which often need a high-temperature process. The nafion layer is stable in ambient environments, without the need of any conformal coating, to the best of our observations to date. We have been successful in demonstrating the use of CBRAM technology for RF switching, using simple fabrication steps.

To aid the understanding of the reader, we reproduce an experiment to prove the ion-conducting properties of nafion for CBRAM/MIM switching applications (Susanti 2016). Here, we fabricate a group of CBRAM/MIM switching cells using a simplified fabrication process, from which the optimized process used by us for all applications in this book, explained later, is originated.

Figure 2.5 presents an example of CBRAM/MIM switches fabricated on a classic copper cladded FR-4 substrate, with copper-nafion-aluminum switching layers. A cross-section view of single copper-nafion-aluminum MIM switch of this batch is shown in the inset of Figure 2.5. The fabrication process of this realization is described below.

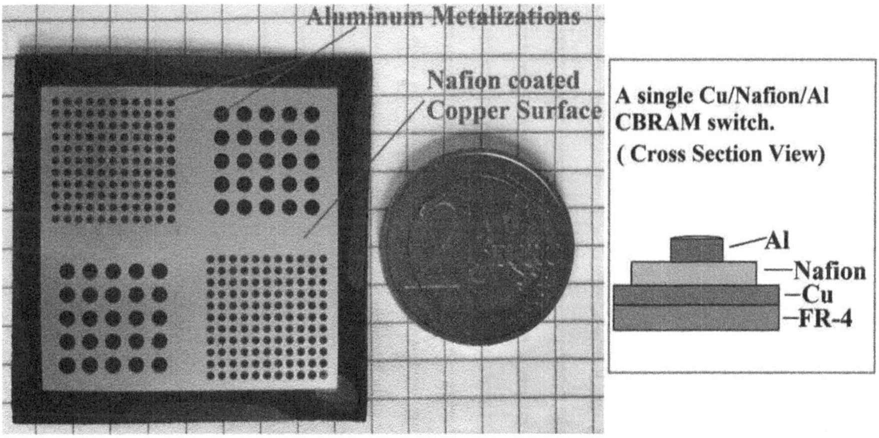

Figure 2.5. *Photograph of copper-nafion-aluminum CBRAM/MIM switch on FR-4 substrate. Inset shows the layers of a single switch. For a color version of this figure, see www.iste.co.uk/jayakrishnan/technology.zip*

First, a piece of copper-cladded FR-4 substrate is chosen and its copper surface is polished mechanically and cleaned with acetone to form a smooth and clean surface. The thickness of this copper clad is specified by the manufacturer as 35 μm. This copper surface acts as the active electrode. Then we spin coat a layer of nafion impregnated resin solution at a rate of 500 RPM for 30 s, and this layer is dried on a hot plate for 60 s. This layer functions as the ion-conductor. The chosen spin coating rate is optimized for a layer thickness of 600 nm, which is observed to give the best switching performance in a previous study targeted on material properties optimization (Susanti 2016). Once the nafion layer is formed on the copper surface, the inert electrode layer, which is aluminum, is deposited using thermal vapor deposition, using a dedicated nickel mask, with an engraved aperture of dots of 1 mm and 2 mm in diameter, as shown in Figure 2.5. The thickness of the deposited aluminum layer is measured to be in the range 300 nm to 1 μm.

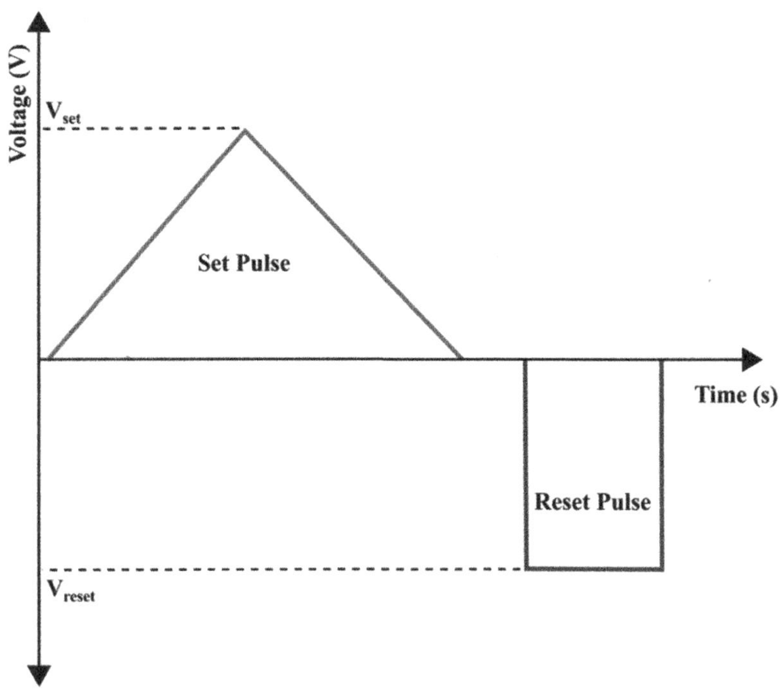

Figure 2.6. *DC pulse waveform used to operate CBRAM/MIM switches*

The presented switches were operated using a current limited voltage wave form generated from a Keithley 2400 source meter using a dedicated Labview GUI. Here, we use a triangular waveform of peak amplitude 12 V starting from 0 V with a compliance current limit of 0.5 mA for set (ON) operation, and a flat square pulse of amplitude –20 V and a compliance current limit of 0.2 A for reset (OFF) operation. Both polarities are applied to the switch with respect to the active (Cu) electrode. Wave shapes of these waveforms are shown in Figure 2.6, and detailed explanations on the operation and selection of this wave shape are given in section 2.2.

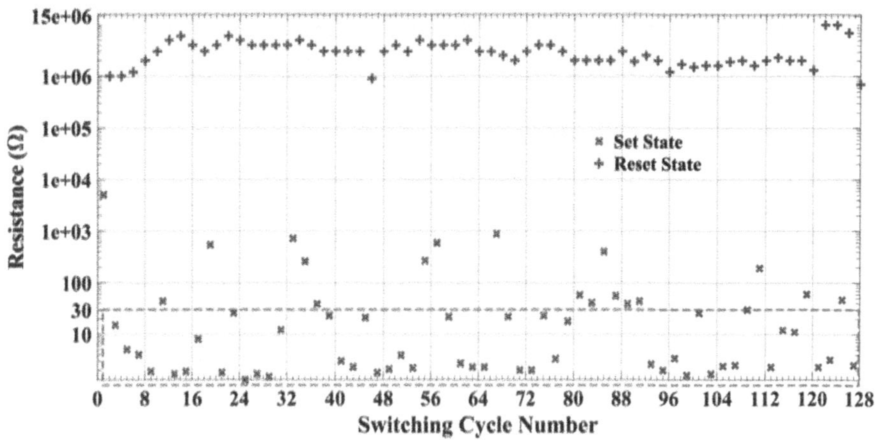

Figure 2.7. *First 130 switching cycles of one typical cell of copper-nafion-aluminum MIM switches shown in Figure 2.5. The dashed-line rectangle shows set states under 30 Ω. For a color version of this figure, see www.iste.co.uk/jayakrishnan/technology.zip*

DC pulses were applied to the switch, carefully probing them on a probe station. Switches were probed individually on the aluminum electrode for each switch, for access to the inert electrode, and on the copper base of the entire structure, after scrubbing off some nafion layer parts in a small area to access the active electrode. Figure 2.7 shows the first 130 switching cycles of a typical MIM switch cell in this batch. We observe that most of the cells in this batch show similar behavior and this is observed also for multiple batches. The majority of the set (ON) state resistances are below 30 Ω, and reset (OFF) state resistances are above 1 MΩ. These set/reset operations were done consecutively in a time period of around 500 s. These results are added here to give information on the switching characteristics of a CBRAM/MIM cell to the reader.

Next, we present in the following sections of this chapter, and in the following chapters, our work and outcomes in the application of nafion-based CBRAM/MIM switches for passive and non-volatile RF devices application.

2.2. CBRAM-based fully passive solid-state RF switch on classic RF substrates: design and process optimization

The primary objective of this research is to optimize the CBRAM switching technology for low-cost RF and microwave devices using a simplified process compatible with mass industrial production, notably utilizing low cost materials which are suitable and stable in ambient environments. In this section, we explain the conceptual validation of use of CBRAM technology for RF switching application within the bounds of the desired objective of this research. Details of design, optimization of proposed fabrication process, and experimental results and discussions of investigations are described comprehensively in the following sections.

2.2.1. *Design of a CBRAM-based shunt mode RF switch*

To validate the concept that CBRAM technology could be used for RF switching, and that it could be fabricated using simple high-scale industrial setup without the help of any clean room equipment, we present the design of a CPW shunt mode RF switch as shown below. The RF switch is designed in shunt mode on a 50 Ω CPW transmission line as shown in Figure 2.8. The switching component – CBRAM or MIM cell – is formed between the 100 μm wide shunt line and the 300 μm wide aluminum metallization sandwiching the 600 nm thick nafion layer, as shown in Figure 2.8. The layer structure of integrated MIM switch is shown with more detail in the cross-section view of the CPW line in Figure 2.9. The shunt line is connected to either ground planes of the CPW and the aluminum metallization is a continuation of the transmission line connecting Ports 1 and 2.

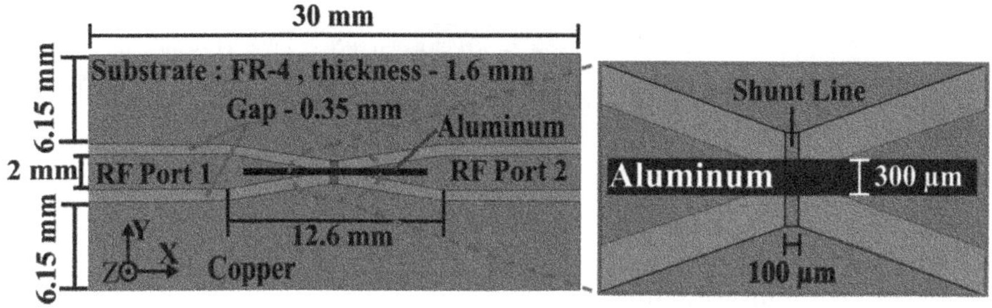

Figure 2.8. *The shunt mode CPW RF switch. The MIM switch is integrated by sandwiching a 600 nm nafion layer among the 300 µm wide aluminum line and 100 µm wide shunt line shown in the zoomed inset. For a color version of this figure, see www.iste.co.uk/jayakrishnan/technology.zip*

The topology presented is initially optimized using full-wave electromagnetic simulation in CST Microwave Studio to choose the most appropriate performances. Conventional materials such as copper, aluminum and FR-4 from the design are modeled using material properties already available with CST Microwave Studio. The ion-conductor layer using nafion is defined as a "new material" with dielectric properties reported in the literature. The dielectric constant and loss tangent of nafion (ion-conductor) is chosen to be 3.5 (dimensionless) and 0.34 (dimensionless), respectively (Lu *et al.* 2009), given for dry nafion membranes. A waveguide port, optimized for 50 Ω line impedance, is modeled to feed RF power to the device. An adaptive meshing scheme is used to optimize the mesh parameters in simulation. Appropriate local mesh is also defined where ever necessary to make sure that the minute thickness of the ion-conductor and other metals are precisely modeled and included in internal calculations of simulation software.

The formation and dissolution of filament is modeled using a lumped resistor with values similar to the measured DC resistance of MIM cells, observed in experiments, like the one given in section 2.1.1, and was then modified with the values obtained in initial experimental trials of the CPW RF switch described here. For the set state, we use 1 Ω and 1 MΩ for reset. Figure 2.10 shows the S-parameter response obtained from simulation studies for the optimized CPW-based shunt mode RF switch illustrated in Figure 2.8. Here, one can see that the chosen geometry of the MIM switch provides a good ON state insertion loss and OFF state isolation as desired for a typical RF switch.

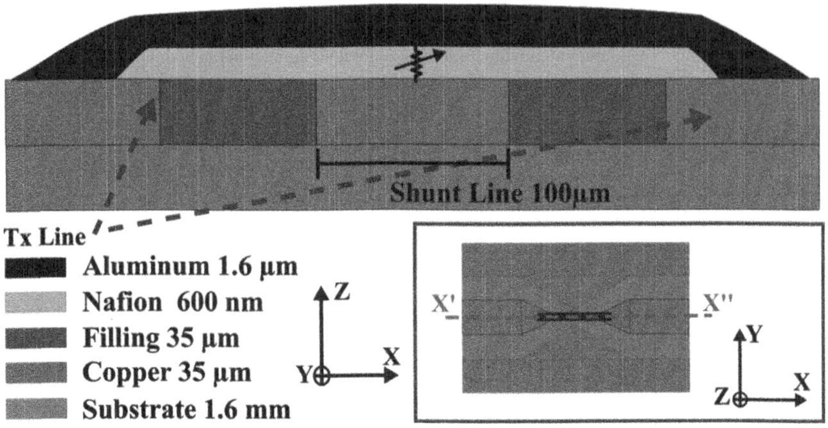

Figure 2.9. *Cross-section view of the CPW shunt mode RF switch showing the MIM structure. Inset shows the plane in cross-section, which is along x' – x'' (please note that the image is an illustrative representation and the layers are not to scale. Please refer to legends for dimensions). For a color version of this figure, see www.iste.co.uk/jayakrishnan/technology.zip*

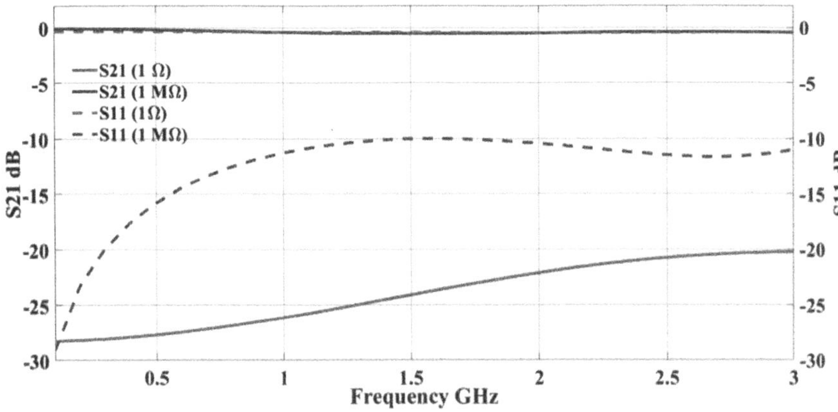

Figure 2.10. *Simulated S-parameters (full wave) of CBRAM/MIM-based 50 Ω CPW shunt mode RF switch shown in Figure 2.8. For a color version of this figure, see www.iste.co.uk/jayakrishnan/technology.zip*

We use a lumped resistor, as explained previously, mimicking the filament resistance instead of modeling a metallic filament to keep the numerical meshing of simulated model in control. A metallic filament, whose dimensions would be typically of the order of a few square nanometers, would require superfine and minute mesh cells to properly resolve it to the internal calculation. Such requirements of very small mesh cells could lead to large overall mesh size and require large computing facilities and an enormous amount of time to complete these calculations. This would eventually result in simulation with reduced accuracy in an ordinary computer used in the laboratory. On the other hand, it is shown below that using the chosen simulation scheme using the lumped resistor explained above gives acceptable results, which are in agreement with experimentally observed characteristics of the fabricated prototype.

The working mechanism of CPW-based shunt mode MIM RF switches could be explained as follows with reference to the geometry depicted in Figures 2.8 and 2.9. Initially, when the MIM switch is in reset or high resistance state, it does not have a significant effect on the RF transmission, and a considerable transfer of power among Ports 1 and 2 can be observed, establishing the RF ON state. For the RF OFF state, upon application of the appropriate switch pulse, a filament is formed between the two electrodes of MIM cell ("set" process). This is similar to a short circuit between a part of the transmission line and ground, and RF power from Port 1 is short circuited to ground, isolating it from the Port 2 (and vice versa), establishing the RF OFF state. Inversely, the connection between the Ports 1 and 2 is resumed when the filament is dissolved by the "reset" process, establishing the RF ON state again, whenever desired.

2.2.2. *Fabrication process*

The fabrication process and associated challenges for proposed low-cost CBRAM RF switches are explained in detail in the following. The substrate chosen for fabrication of the switch is a classic low-cost FR4 epoxy, commonly used for general purpose electronics and RF circuit prototyping. Initially, the copper pattern of the CPW line given in Figure 2.8 (except the nafion layer and aluminum metallization) is engraved on the substrate through classical UV lithography. Patterned copper on the substrate typically has a thickness around 35 μm. This thickness is a good parameter in terms of skin depth for a good RF conductivity and transmission at the chosen frequencies. However, it is an undesirable factor and would lead to some very fragile soft spots in the upcoming layers that will deploy nafion and aluminum in the following fabrication steps. The deposited nafion and aluminum layers are conformal to the shape of copper pattern, and their thicknesses are very low (<2 μm) in comparison to the large step (~35 μm) due to copper thickness. This would tend to create some frail edges in the nafion and aluminum layers that could eventually break due to small mechanical shocks or vibrations, or thermal effects due to operating DC pulses or RF power, and hinder the desired functioning of the switch. A pictorial illustration of the cross-section view of such possible weak spots due to the thickness mismatch of associated materials layers is shown in Figure 2.11. Such an associated challenge is identified in the initial prototypes of fabricated design and is solved as follows.

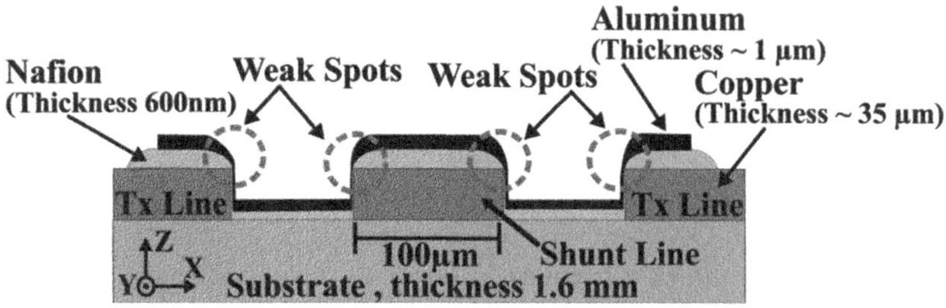

Figure 2.11. *Possible fragile soft spots and step discontinuity due to thickness mismatch among copper nafion and aluminum layers constituting the CBRAM/MIM switch. For a color version of this figure, see www.iste.co.uk/jayakrishnan/technology.zip*

To mitigate this issue, this step discontinuity is filled using a synthetic epoxy resin and mechanically polished to reveal the copper surface. This process ensures a smooth finish that helps to achieve even thickness for the electrolyte and metal layer, as shown in Figure 2.9. Then the substrate is heated on a hot plate at 100°C for 4 min to build a thin layer of copper oxide on the surface, which is found to enhance the efficiency of reset cycles from observations and experience from previous studies. The nafion (ion conductor) layer is then formed by spin coating the "nafion perflourinated resin solution" (Sigma-Aldrich 2019), at 500 rpm for 30 s, followed by heating the substrate at 100°C for 60 s, to dry the layer. In our experience, it is seen that the spin-coated nafion layer is completely dry in a few minutes in ambient air. However, the drying process on a hot plate is adopted to speed up the process and ensure better quality of fabricated samples. Then the aluminum metallization that connects the two copper segments of transmission line, which originates, respectively, from Ports 1 and 2, is deposited by the thermal evaporation technique, with the help of a dedicated Nickel pattern mask. This mask carries an aperture with the dimensions of the metallization, as shown in Figure 2.8.

The thickness of the nafion layer and aluminum metallization are measured to be 600 nm and 1.6 μm, respectively, using a Dektak-150 mechanical profilometer in the presented fabrication. Generally, it is seen that the thickness of aluminum using this process varies in the range of 300 nm–1.6 μm for different batches and is observed to exhibit considerable performances. A photograph of a fabricated CBRAM-based CPW shunt mode RF switch, ready for experimental characterizations, is shown in Figure 2.12.

Figure 2.12. *Photograph of the fabricated switch showing the microphotograph of the MIM switch area in the inset. For a color version of this figure, see www.iste.co.uk/jayakrishnan/ technology.zip*

This fabrication process explained above is evolved experimentally after numerous modifications through trial and error over time. This process is currently used as the standard method for fabrication of CBRAM/MIM switches on rigid substrates within our group and is utilized for all fabrications of similar type presented in this book, unless otherwise mentioned.

DC operating pulses for switching the device are generated using a Keithley 2400 source meter, controlled using a dedicated Labview GUI. DC triangular pulses of peak amplitude 18 V with a compliance current limit of 0.5 mA and flat voltage pulse of peak voltage –20 V with a compliance current limit of 0.2 A are used, respectively, to set (ON) and reset (OFF) the CBRAM switch. Voltage and current plots of the switching pulse during a typical switching operation are depicted in Figures 2.13 and 2.14, respectively. The plot in Figure 2.14 also shows the evolution of switch resistance with time.

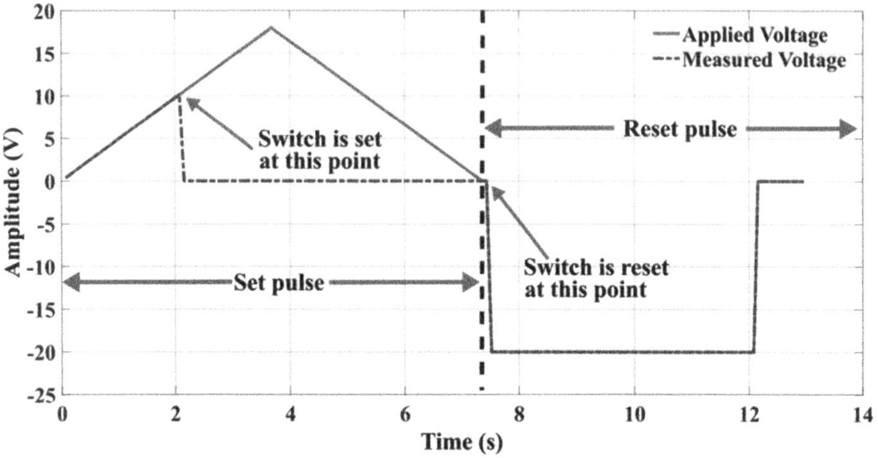

Figure 2.13. *DC voltage waveform used for operating the switch. The switch is initially reset up to 2 s, set from 2 s till 7.75 s, and finally reset from 7.75 s until 14 s. For a color version of this figure, see www.iste.co.uk/jayakrishnan/technology.zip*

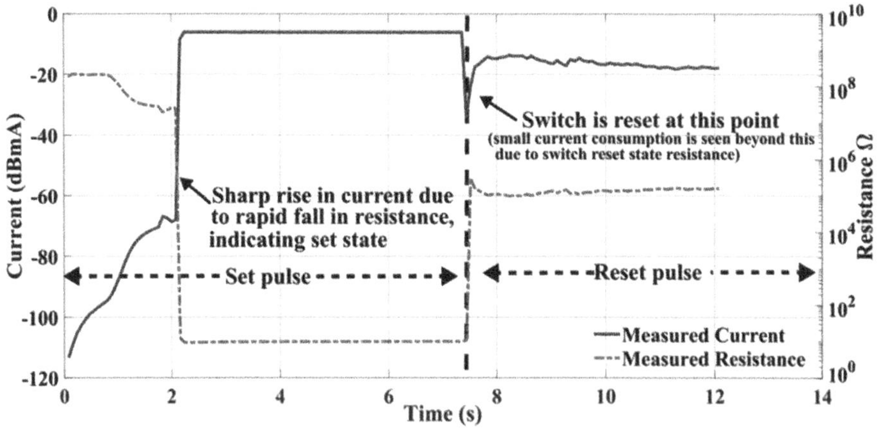

Figure 2.14. *DC current waveform measured along the current path of switch during set/reset operation, along with time evolution of switch resistance. The switch is initially reset up to 2 s, set from 2 s till 7.75 s and finally reset from 7.75 s until 14 s. For a color version of this figure, see www.iste.co.uk/jayakrishnan/technology.zip*

In the presented case in Figures 2.13 and 2.14, the CBRAM switch is initially reset up to 2 s, set from 2 s till 7.75 s, and finally reset from 7.75 s until 14 s and beyond. Here, the CBRAM switch is set at 10 V (around 2 s in time scale), marked by an abrupt fall in voltage across the switch and a sharp rise in current, acknowledging the lowering of the impedance state due to filament formation. It is reset instantaneously at –20 V (around 7.75 s), exhibiting a sharp rise in resistance across the switch, showing appropriate voltage drop across it.

Readers should not be confused with the minute current consumption seen across the switch after reset state (after 7.75 s). This is a dissipation due to the high impedance state of the switch and is seen to be higher than dissipation at the high impedance state before set state (from 0 to 2 s) since the compliance current limit for reset pulse (0.2 A) is at least 400 times greater than the limit for set pulse (0.5 mA).

Pulse shape of the applied voltage pulse is plotted in a pink continuous line in Figure 2.13. The measured voltage plotted in the same figure in a blue dashed line follows this shape in cases where the set operation is not successful and the switch retains a high impedance state. Instantaneous values of power used for set and reset are calculated from values of the voltage and current plotted in Figures 2.13 and 2.14, and are found to be, respectively, 1.7 μW and 3 mW for the set and reset processes. DC switching pulses were fed to the device initially by probing the CBRAM/MIM cell using a microprobe station, but later this was done directly through the RF connectors (removing the RF feed cables each time). DC operating pulses could also be fed to the switch along with the RF power if necessary, for a continuous operation, employing proper isolation circuitry. These presented DC operating pulse waveforms are used as the standard method for switching CBRAM/MIM switches in this study, unless otherwise mentioned.

2.2.3. *Results and discussions*

Radio frequency characterizations of fabricated devices were carried out using the Agilent ENA E5061B (3 GHz Network Analyzer), using SOLT calibration, and with a power level of 0 dBm. The switch shows good performance up to 3 GHz, and the results are given in Figures 2.15 and 2.16, along with full-wave electromagnetic simulation results for comparison.

Experimental observations given in this section are inclusive of the effects of transmission lines and connectors feeding the switch.

Here, one can see an RF ON (reset) state insertion loss of 0.5 dB at 1 GHz and lower than 1.1 dB up to 3 GHz. In the RF-OFF (set) state, the isolation is greater than 16 dB till 3 GHz. Readers should not be confused here by the terms RF ON and RF OFF, which are applicable only to the presented complete model of the RF switch and set (ON) and reset (OFF) states, applicable only to the integrated CBRAM switch. One should keep in mind that the CBRAM switch is integrated and operates, in the CPW RF switch, in shunt mode as explained in section 2.2.1.

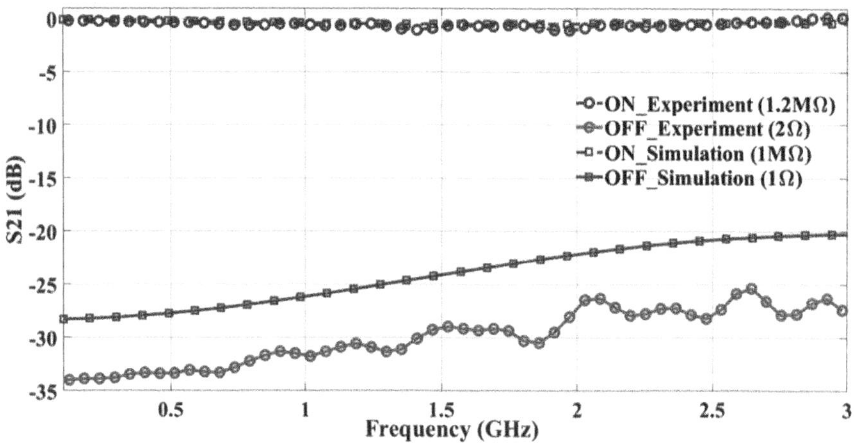

Figure 2.15. *Experimentally obtained S-parameters of CBRAM-based CPW shunt mode RF switch: S21. DC resistance of each state is given in brackets of corresponding legends. For a color version of this figure, see www.iste.co.uk/jayakrishnan/technology.zip*

DC filament resistances of the CBRAM switch measured during the experiment using the Keithley 2400 source meter are found to be in the range of 2 Ω to a few tens of ohms, and hundreds of kΩ to MΩ in the set (R_{SET}) and reset (R_{RESET}) states, respectively, for different switching cycles. On average, the experimentally observed switching cycles show a maximum R_{SET} of 5 Ω and minimum R_{RESET} of 1 MΩ, which is a good R_{RESET}/R_{SET} ratio.

Table 2.1 summarizes and compares the performance of the presented CBRAM-based CPW shunt mode RF switch to state-of-the-art switching devices in the field. Even though the presented RF switch is a budding technology in its infancy stage, and not mature enough to be compared to the state-of-the-art, this information gives an insight into the capabilities and relevance of this concept.

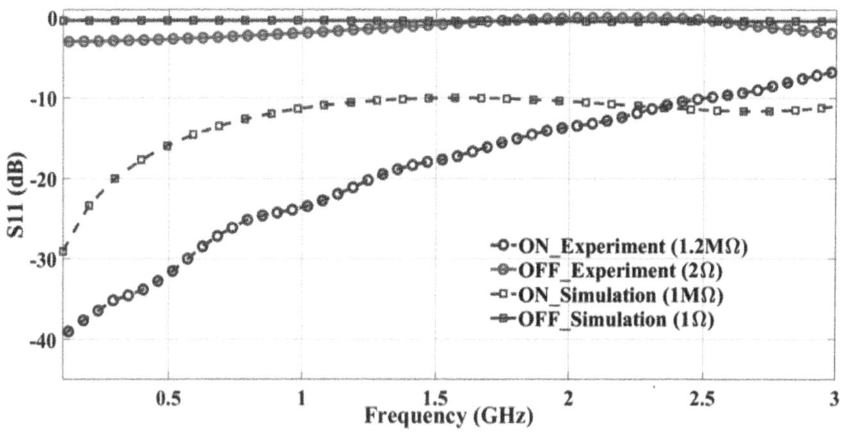

Figure 2.16. *Experimentally obtained S-parameters of CBRAM-based CPW shunt mode RF switch: S11. DC resistance of each state is given in brackets of corresponding legends. For a color version of this figure, see www.iste.co.uk/jayakrishnan/technology.zip*

From Table 2.1 one can see that the presented CBRAM-based shunt mode RF switch has considerable entry level performances, which could be expected from the chosen fabrication process. The minimum feature size requirement for the proposed switch is a great achievement among the choice of fabrication process and its desired requirement. The expected fabrication tolerance of this method from the standard process used in state of the art devices in this aspect is at least 10 times enhanced. This gives one a good flexibility in the mask or layers alignment accuracy required for microfabrication and allows these realization processes to be carried out in an ambient room rather than in a clean room environment. Moreover, the proposed fabrication process could be further optimized for mass production in an industrial environment for microfabrication. This would be without any complicated requirements like in a clean room and would contribute to a reduced cost factor from a commercial point of view, which is a promising future aspect of the presented technology. In these experiments, we can also see that the fabricated devices are seen to exhibit good reproducibility and switching reliability.

Features	**Our design**	**Nanoionic (CBRAM) RF switches state of the art (Nessel and Lee 2011)**	**MEMS RF switches (Goldsmith *et al.* 1998)**
"Set" (ON) state resistance (average)	2–5 Ω	10 Ω	0.35 Ω
Actuation voltage range	~10 V/–20 V#	~1 V	30–50 V
RF insertion loss (average)	0.5 dB	0.5 dB	0.25 dB
RF isolation (average)	-30dB	-35dB	-35 dB
Operating frequency range	DC to 3 GHz	DC to 6 GHz	DC to 35 GHz
Switching speed	*	1–10 μs	~5 μs
Minimum feature size (resolution) requirement	100 μm	10 μm	<1 μm
Alignment accuracy required for fabrication	Low	Very high	Very high
Fabrication environment	Ambient room	Clean room	Clean room
Power consumption (to change state)	μW/mW#	μW	μW
Power consumption (to maintain state)	0	0	μW
Fabrication cost (laboratory perspective)	Low	High	High
Manufacturing cost (industrial perspective)	Low	Moderate	High

*At present, we do not have enough information for calculating the switching speed of the presented CBRAM devices. In order to ensure stable and smooth switching, a gradual voltage sweep is used to operate the switch (as shown in Figures 2.13 and 2.14) and this limits us in determining the switching speed. However, from experiments, it is clear that the switch can be operated using instantaneous voltage pulses. Such short pulses could be optimized to operate the switch after a detailed study of the DC dynamics, after which one could accurately calculate the switching speed.

#For set and reset process, respectively.

Table 2.1. *Comparison of realized CBRAM-based CPW shunt mode RF switch with the state of the art design of RF switches*

A batch of fabrication of the presented design consists of around two to four switches and most of the time all of the devices are seen to be functional, except some minor defects in the RF performance due to fabrication tolerances.

2.3. Electrical equivalent model analysis

An approximated electrical lumped-parameters circuit model for a CBRAM unit cell switch is depicted in Figure 2.1(d). The presented model consists of a filament resistance R_{MIM} and an equivalence cell capacitance C_{MIM} of a chosen CBRAM switch. This electrical model is chosen based on the geometric arrangement of a CBRAM/MIM cell, and is a commonly used approximation in recently reported CBRAM based RF switches in the literature (Pi *et al.* 2015).

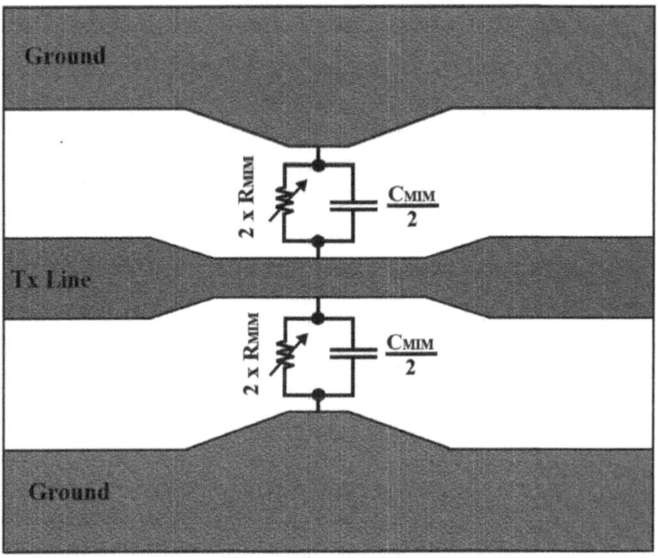

Figure 2.17. *Electrical equivalent model of CBRAM/MIM switch attached to a CPW line*

Here, we present an electrical model of the presented CBRAM-based shunt mode RF switch and more specifically explain the functioning of the integrated CBRAM cell in the presented switch through an equivalent model representation given in Figure 2.17. Validation of this model is done in full-wave electromagnetic simulation as follows. The RC parallel model of the CBRAM switch is inserted into a CST simulation model of the proposed CPW shunt mode RF switch as follows. Initially the ion-conductor layer and shunt line from the design of the presented switch (given in Figure 2.8) is removed, and then the RC–parallel circuit (Figure 2.1(d)) of the CBRAM switch is added in its place, as shown in Figure 2.17, as discrete components.

Here, the RC parallel circuit is distributed symmetrically on both sides of the CPW transmission line for obtaining a homogenous current distribution. This is similar in effect to adding a discrete MIM switch to a normal CPW transmission line tapered at the center.

Parametric values of this electrical model are optimized by fitting simulated RF characteristics with experimentally measured results. Values of resistance R_{MIM} that fit experiments are 1 Ω and 1 MΩ, for set (RF OFF) and reset (RF ON), respectively, and for C_{MIM} 1.55 pF. This resistance value is similar to the measured DC resistance value of filament impedances (~2–5 Ω for set and ~1–2 MΩ for reset) in experiments. The capacitance (C_{MIM}) value is similar to that calculated from the geometry of the 100 μm × 300 μm CBRAM/MIM cell (see Figures 2.8 and 2.9) filled with nafion as ion conductor, through standard equations using reported dielectric properties of nafion (Lu *et al.* 2009), and is found to be 1.55 pF. These values are summarized in Table 2.2. Figures 2.18 and 2.19 show the S-parameter values of the simulated (full-wave) electrical equivalent model explained here along with experimental results for comparison. Considerable agreement among the results of the electrical model and experiment given in these plots validates the accuracy of the proposed concept.

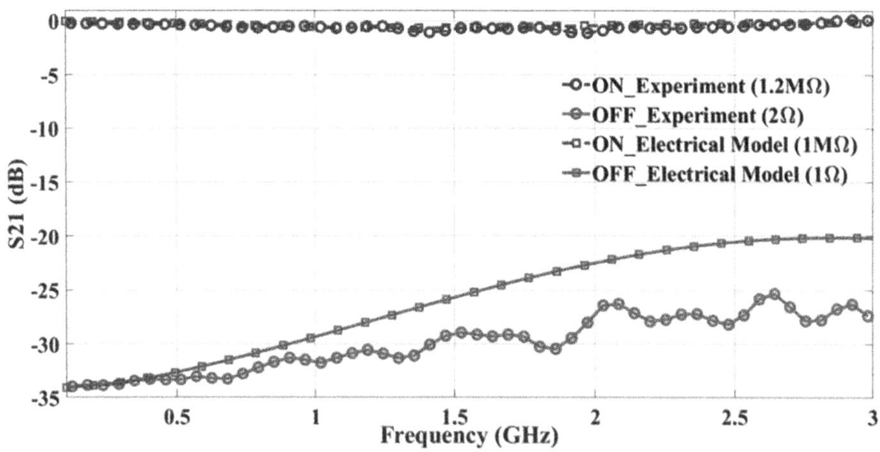

Figure 2.18. *S21 characteristics of the electrical equivalent model of the CBRAM-based CPW shunt mode RF switch given in Figure 2.17, along with experimentally obtained data for comparison. For a color version of this figure, see www.iste.co.uk/jayakrishnan/technology.zip*

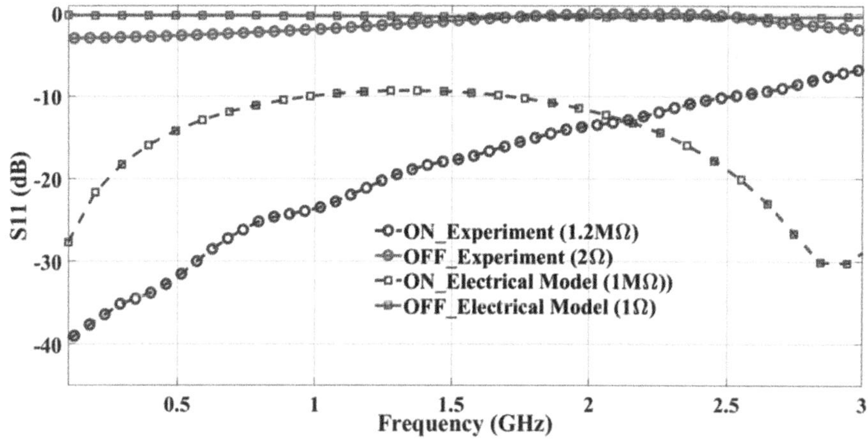

Figure 2.19. *S11 characteristics of electrical equivalent model of the CBRAM-based CPW shunt mode RF switch given in Figure 2.17, along with experimentally obtained data for comparison. For a color version of this figure, see www.iste.co.uk/jayakrishnan/technology.zip*

Parameters	Electrical model		Experimental results (R_{MIM}) Calculation (C_{MIM})	
	Set	Reset	Set	Reset
R_{MIM}	1 Ω	1 MΩ	2–5 Ω	1–2 MΩ
C_{MIM}	1.55 pF	1.55 pF	1.55 pF	1.55 pF

Table 2.2. *Parameter values of analytically fitted electrical model, along with experimentally calculated values for the CBRAM-based 50 Ω CPW shunt mode RF switch*

2.4. Effect of filament resistance of CBRAM switches on RF transmission

Filament resistance (R_{MIM}) of a CBRAM/MIM cell to some extent is a function of the voltage and current used for set (ON) and reset (OFF) process (Pi *et al.* 2015), as depicted in Figures 2.13 and 2.14 for this experiment. However, there are not much information available on controlling the filament resistance for a desired need, at this point of time. At present, optimization of the DC waveform is undertaken to achieve a filament resistance value below and above a given threshold, respectively, for the set and reset states, for example, above a few kΩ for reset, and below 20 Ω for set states, to ensure a good RF transmission and isolation characteristic as discussed for the RF switch above.

Here, we present a study to show the effect of filament resistance on transmission characteristics of the presented CBRAM-based CPW shunt mode RF switch. For this, we use the same model of shunt mode CPW RF switch explained in section 2.2 and simulate for different values of filament resistance of integrated CBRAM switch. The results of this study are depicted in Figure 2.20.

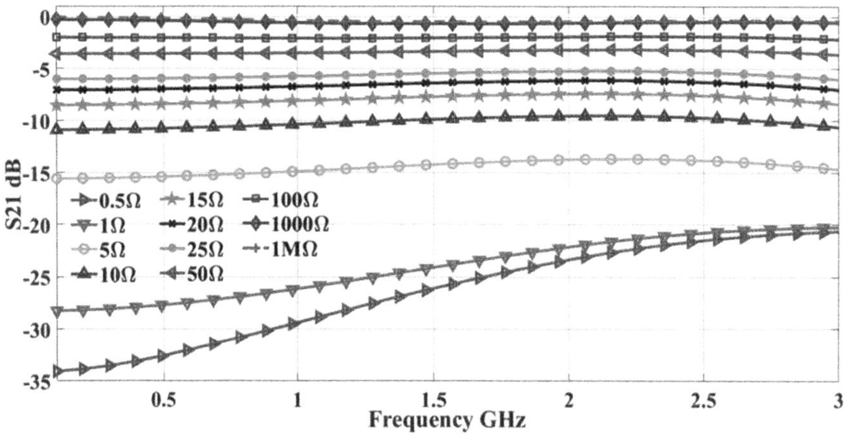

Figure 2.20. *Effect of filament resistance of CBRAM/MIM cell on RF transmission. Full-wave simulation studies on CPW shunt mode RF switch given in Figure 2.8. For a color version of this figure, see www.iste.co.uk/jayakrishnan/technology.zip*

The switch shows an acceptable RF ON state performance for filament resistance values above 1000 Ω of the CBRAM cell normally identified as the reset state. Good RF isolation characteristics defining the RF OFF state are exhibited by the switch for filament resistance values of the CBRAM cell below 10 Ω identified in general as the set state. In addition, it is interesting to note that we can see from the outcome of this study in Figure 2.20 that it is possible to define several intermediate RF transmission values by effectively controlling the filament resistance values of the integrated CBRAM switch. This quantization of RF transmission levels effectively reveals that the potential ability of the integrated CBRAM cell in this RF switch can function as an electronically programmable RF attenuator.

Programmable attenuation characteristics in the presented CPW shunt mode RF switch were demonstrated experimentally for the model depicted in section 2.2.1 and are given in Figure 2.21. This experiment was carried out by programming (set and reset process) the integrated CBRAM switch through several trial and error attempts by controlling set and reset voltage pulses, till desired impedance was achieved.

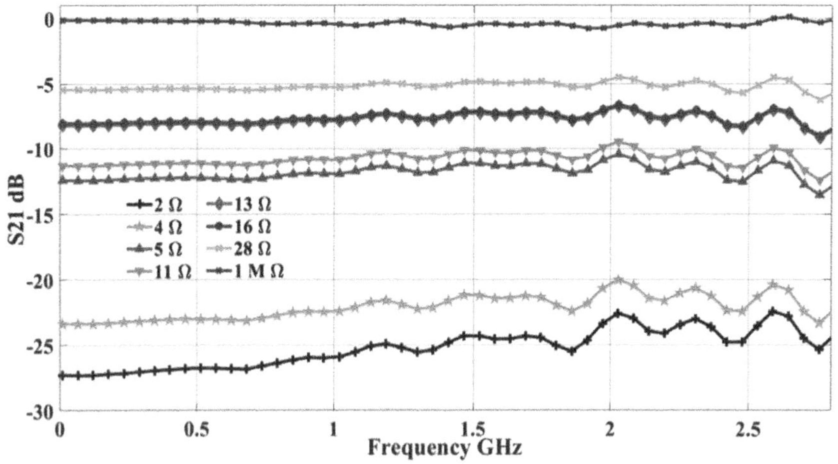

Figure 2.21. *Experimentally observed RF attenuation characteristics for different filament resistance of integrated CBRAM/MIM cell in CPW shunt mode RF switch (design shown in Figure 2.8). For a color version of this figure, see www.iste.co.uk/jayakrishnan/technology.zip*

Such CBRAM switches could be configured in a desired configuration (e.g. "T", "π", or hybrid topologies) for a required programmable attenuator need. With the help of an efficient strategy to properly predict and achieve a desired filament resistance for CBRAM cells, we could envision a promising future for such fully passive and non-volatile electronically programmable RF attenuator components.

The proposed design of electronically programmable and non-volatile RF attenuators could find application in a variety of RF and microwave solutions that demand a low power budget such as mobile communication equipment, satellite applications and so on. Techniques for control of filament resistance of CBRAM cells and development of programmable RF attenuators, and similar devices based on this principle, are in progress and are expected to be revealed soon in the literature.

2.5. Time stability, switching cycles and other interesting features

In this section, we discuss the time stability of set (ON) and reset (OFF) states of CBRAM switches, and the maximum achievable number of switching cycles using devices realized through the proposed fabrication technique. Time stability or impedance retention time of a non-volatile switch is an important aspect, which defines the reliability of the device. This is defined as the maximum amount of time the switch would be able to retain its impedance state, after being latched to a particular impedance (set or reset state), without requiring any actuation or power supply.

Practically, it is obvious that no non-volatile switch, unless it is a mechanical latch, provides infinite impedance retention. Yet, time stability of an achieved impedance state for a considerably long duration, which could fulfill the requirements with respect to the desired application, is an acceptable feature. In such a case, the switch state could be refreshed at periodic intervals to maintain a given state, thereby lowering the power budget of target application.

Figure 2.22 shows the outcome of an experiment to validate the stability of proposed CBRAM/MIM RF switch states over time. Here, we establish the set state of a typical MIM switch, using the DC pulses as explained above, and then measure the impedance state of the switch in regular intervals. The impedance measurement is done using the Keithley 2400 source meter, with a voltage and current level much smaller than the set and reset pulse (1 V, 50 μA) for a time period exceeding 1000 min. Then the switch was reset and the same experiment was carried out in this state also. The outcome of this experiment is given in Figure 2.22. From Figure 2.22, it is apparent that the presented switch is stable for more than 1000 min in both set (ON) and reset (OFF) states. A detailed study on the time stability of the proposed device has not yet been carried out as it is beyond the current scope of this research activity, which is to prove the feasibility of CBRAM technology for RF switching applications, using simple fabrication techniques. However, this is one of the prime future goals of this research. Moreover, state of the art devices reported among RF switches based on CBRAM technology claims an impedance retention time better than 3.3 years (Pi *et al.* 2015), which is a very good number at such an early stage of this technology. Further optimizations in material science aspects in this field are expected to yield better results in terms of impedance retention and time stability of CBRAM switches in the coming years.

Next, we discuss the maximum number of switching cycles possible with CBRAM RF switches, realized using the proposed fabrication process in this chapter. The number of achievable switching cycles is an important feature of a switching technology.

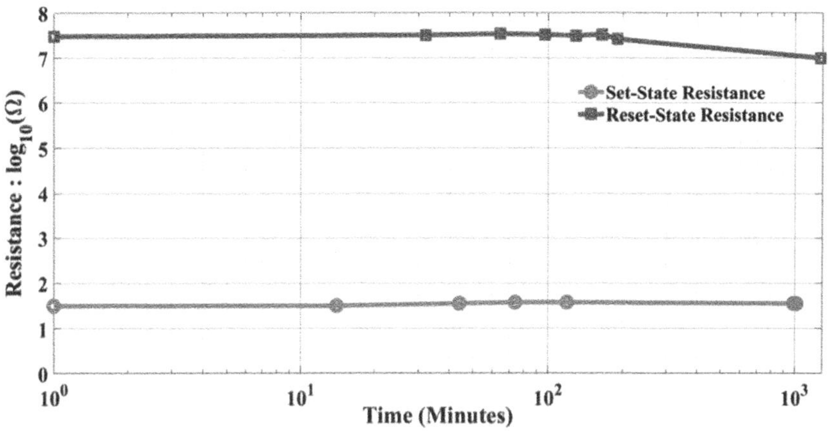

Figure 2.22. *Time stability of set/reset states of a CBRAM/MIM switch*

An experiment described in Jian *et al.* (2017) details similar CBRAM switches, realized using the fabrication process described in this chapter, which exhibit more than 2,000 cycles of continuous operation. This is considered to be a good number of operating cycles using such a simplified fabrication process in an ambient environment without any “clean room” processes. It is true that the number of switching cycles is only one of the performance factors for commercially used proven technologies in this field like semiconductor-based RF switches. Nevertheless, CBRAM switches used for RF applications, with continuous switching cycle repeatability, limited to around 2,000 cycles, could still cater to the needs of low-cost passive RF applications like a disposable transport ticket based on an electronically reconfigurable chipless RFID tag (explained in detail in Chapter 3) to mention just one. At present, with reference to the literature on the reliability of this technology (Jian *et al.* 2017; Adesto Technologies 2016), CBRAM/MIM switches show promising performances that could cause them to emerge as the new trend of RF switching in the near future.

Detailed investigations on the maximum achievable number of switching cycles and optimization of material and design aspects in this domain are dedicated specialized topics, given the utmost importance in continuing research in this domain.

In all the applications of CBRAM RF switches described in this book, as a quality factor benchmark, the presented switches were tested for more than 50 cycles of operations and most devices were retained as fully functional after this test. This is a good number of operation cycles for the nearest application targets of these devices, like in electronically rewritable chipless RFID labels, as indicated above, and similar devices.

The proposed switch is believed to be mechanically robust due to the fact that it does not have any mechanical or moving parts, which would be damaged as a result of any vibrations or shocks. All layers integrated into the device are conformal to each other and would contribute to a mechanical robustness better than switching solutions like MEMS. Additionally, the switches are integrated into the target device inherently in the design itself. This is a benefit in terms of saving space and reducing perturbations in the device, as required for classic switch solutions, like conductive gluing or soldering.

A multidisciplinary performance aspect of the proposed solution is the well-known humidity sensitivity properties of nafion (Wu *et al.* 2006; Liu *et al.* 2018). Nafion is a renowned humidity sensor, and this quality could be a contradictory argument for the current application. The humidity sensitivity of nafion may derail the desired functioning of the switch, depending on variations in atmospheric humidity conditions, contributing to malfunctions in the desired filament impedances and switching speeds. However, we would like to challenge such an argument with the fact that in our experience, the switches were seen to exhibit repeatable results even when tested after 6 months from manufacture. In addition, the nafion layer used in the switch is naturally covered with a conformal metallic layer, as shown in Figures 2.8 and 2.9. In spite of that, it is possible to provide a conformal protective coating for the MIM switch area on the device to protect it from the ambient environment (Havens *et al.* 2002). This would solve the issue of atmospheric interference on the desired functioning of the device.

Another interesting observation about the amount of water content in nafion is as follows. Ion conductivity of nafion is directly proportional to its water content (Heitner-Wirguin 1996). Increased ion conductivity or increased water content could thus help us with MIM switching at a greater speed and with greater ion conductor thickness. Greater thickness of ion conductors is an interesting factor for RF application, as it would help in reduction of reset (OFF) state capacitance of a typical MIM cell. In an experiment related to observation of filament formation in CBRAM/MIM switches using nafion, we observed this dependence of the switch gap on the nafion layer's water content. We were able to grow filaments in a switching gap, which is about 100 μm wide, in place of the 600-nm thick ion conductor layer in the presented case. This study is not given in this section as it is not complete, and has no direct connection with RF switching applications at present, and moreover the switch topology and probing methods used are different from the current methodology. The photographs and details of this experiment are given in the Appendix.

However, the expectation of reduced capacitances due to thicker ion conductor layers is contrary to the fact that higher water content would lead to higher dielectric constants, and dielectric losses, due to water molecules. A dedicated and detailed experimental study in this regard is in progress, aimed at optimization of the best achievable RF switching characteristic and will be published soon in near future.

2.5.1. *Reason for choice of CPW transmission line for presented switch*

In the presented model of the CBRAM-based RF switch in section 2.2, a CPW shunt mode is chosen over a classical microstrip transmission line design, in which two simple microstrip segments sandwich the ion conductor layer. This appears to be simpler in terms of design and fabrication than the presented topology using a CPW line.

Despite this, the CPW line is finally chosen in this realization due to limitations imposed by the mechanical resolution available with the proposed realization process, as explained below. The choice of a microstrip design as shown in Figure 2.23 is limited by the considerably large capacitance of CBRAM switches (C_{MIM} ~1–2 pF) at the frequencies of interest (1–5 GHz). This would lead to poor switching behavior and would demand a choice of very small dimensions of electrode areas for lowering OFF state capacitance values to achieve the same performance as with the CPW line.

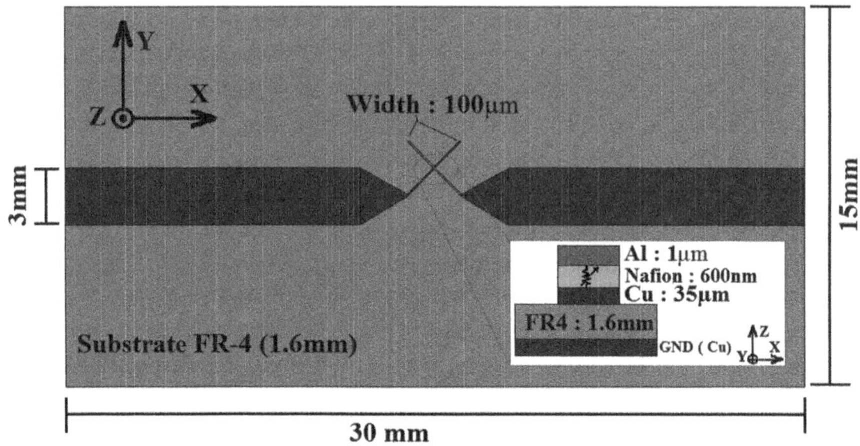

Figure 2.23. *Design of a CBRAM-based RF switch using microstrip transmission line. For a color version of this figure, see www.iste.co.uk/jayakrishnan/technology.zip*

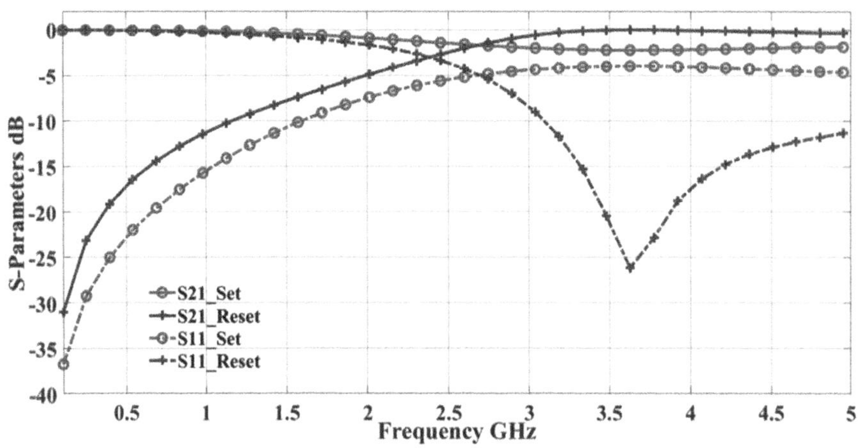

Figure 2.24. *Simulated S-parameters of CBRAM/MIM-based RF switch on microstrip line. For a color version of this figure, see www.iste.co.uk/jayakrishnan/technology.zip*

The result of the full-wave electromagnetic simulation of the RF switch in microstrip design in Figure 2.23 is given in Figure 2.24. It is apparent from this outcome that the RF switching performance is limited to 1.2 GHz, even with an electrode width of 100 μm, which is almost the lower limit of the feature size of the fabrication facility used for the proposed technique in this chapter (fabrication in ambient laboratory without clean room support).

On the other hand, such a design could be made possible by utilizing CBRAM/MIM cell structures with lower OFF state capacitance. This could be achieved either by using planar switch designs as in Nessel and Lee (2011) and Pi *et al.* (2015), which have reduced electrode surfaces and hence provide low capacitances, or by using a higher thickness of the ion conductor layer as discussed in the above section. Another possible technique is to add more switches in series, which could reduce the net capacitance of the microstrip line facilitating better upper frequency limits. Investigations of these aspects are in progress with the authors and are expected to provide devices with improved performance in the near future.

2.6. Fabrication technique for realization of CBRAM/MIM RF switches on flexible substrates

The requirement for an accurate and precise geometric feature control is an uncompromised necessity for most microwave engineering-based products. This requires a sophisticated production setup, which adds to the cost factors of the final product. The cost of production of any device can be brought down, to some extent, by utilizing mass production through an automated setup. In this section, we propose a simple method for realization of CBRAM/MIM-based RF switches. This technique is suited for all types of substrates and especially for flexible substrates like paper. Moreover, this method is compatible with an automated industrial process for mass production.

2.6.1. *CBRAM-based fully passive solid-state RF switch on flexible paper substrates*

In this section, we demonstrate the fabrication of a CBRAM-based CPW shunt mode RF switch on a paper substrate. For this, we chose 80 g/m^2 (GSM) glossy paper. This paper is flexible and has an improved surface for better adhesion of metal layers. The optimized fabrication process for this fabrication is explained as follows.

We demonstrate the realization of a CBRAM-based CPW shunt mode RF switch similar to that given earlier in Figure 2.8, however this time on a flexible paper substrate. The modified layer structure of this RF switch, matched to the flexible paper substrate and its cross-section view, is depicted in Figure 2.25. Here, the design is more simplified, without the requirement of an intermediate filling and polishing process as explained in section 2.2, as all the CBRAM cell layers are of similar thicknesses and are conformal to each other.

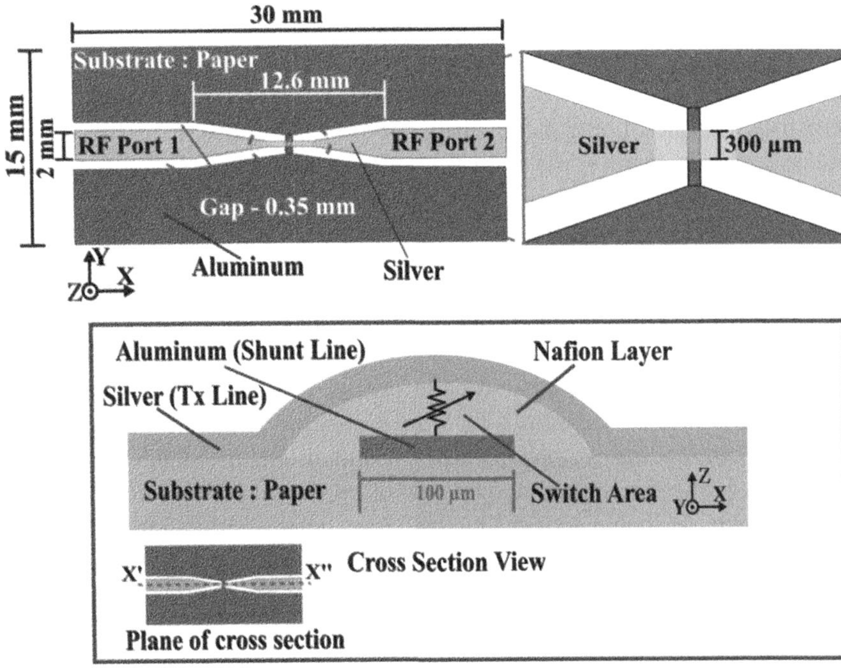

Figure 2.25. *CBRAM-based CPW shunt mode RF switch on paper substrate. Inset in blue border shows the cross-section view of the RF switch in the plane x' – x". Substrate: 80 GSM glossy paper; layer thickness: aluminum: 1 µm, silver: 1.3 µm, nafion: 600 nm. Please note that this image is only an illustration and is not to scale. Kindly refer to legends and explanations for dimensions. For a color version of this figure, see www.iste.co.uk/jayakrishnan/technology.zip*

Silver and aluminum are used here to form the active and inert electrodes, respectively. Nafion is used as the ion conductor. Metal layers of CBRAM electrodes are formed on the substrate by thermal vapor deposition of silver (active electrode) and aluminum (inert electrode). Thermal vapor deposition is done with the help of dedicated nickel masks on which an aperture of the presented topology in Figure 2.25 is engraved. Initially one of the electrodes, which could be silver or aluminum, is deposited in this way. After formation of one of the electrodes, the ion conductor layer is formed by spin coating the nafion solution (Sigma-Aldrich 2019) at a rate of 500 RPM for 30 s for a layer thickness of 600 nm. This nafion layer is then dried on a hot plate at 100°C for 2 min.

We should note here that the drying process on a hot plate is done to save time. This layer could be naturally dried at room temperature with an exposure of about a few minutes. The opposite electrode is then formed using thermal vapor deposition of the appropriate metal. Screw-on SMA connectors are then added to the device to prepare it for experimental characterizations. The whole process consists of only three major steps and is summarized pictorially in Figure 2.26: step 1, thermal vapor deposition of one of the electrode; step 2, spin coating of ion conductor; step 3, deposition of opposite electrode.

The proposed CBRAM/MIM switch could be integrated into most RF circuits using the three-step process as explained herewith. These presented steps could be easily optimized to a roll-to-roll process carried out using an automated industrial setup like that used for manufacture of aluminum-coated PET food packages. Moreover, the use of silver could be limited to the switch area only, by optimizing the nickel mask design, so as to further reduce the cost of realization. The spin coating method, which is not so friendly with a roll-to-roll process, could be replaced by a similar technique such as slot-die coating, dip coating or spray coating for better automation of proposed fabrication process. Potential upgrades in the roll-to-roll process could involves be the replacement of thermal vapor deposition techniques for realization of metallic layers, either fully or partially, with techniques that offer better spatial resolution, accuracy and speed, like inkjet printing.

A photograph of the realized CBRAM-based CPW shunt mode RF switch is shown in Figure 2.27, whose inset shows a zoomed microphotograph of the MIM switch area, revealing the sandwich structure. The layer thickness of silver and aluminum deposit is measured to be 1.3 μm and 1 μm, respectively. Similarly, the nafion deposit has a thickness of 600 nm. These measurements are done with the help of Dektak-150 mechanical Profilometer. Please note that the position of silver and aluminum layers are interchangeable (one on top of other sandwiching nafion layer), provided the switching pulses are applied with correct polarity. Possible combinations in this case (silver on top of aluminum, or aluminum on top of silver) are different only in the area where CBRAM sandwich is formed, and no difference is found in their performance characteristics. Experimental characterizations of and results concerning these switches are presented in detail in the following sections.

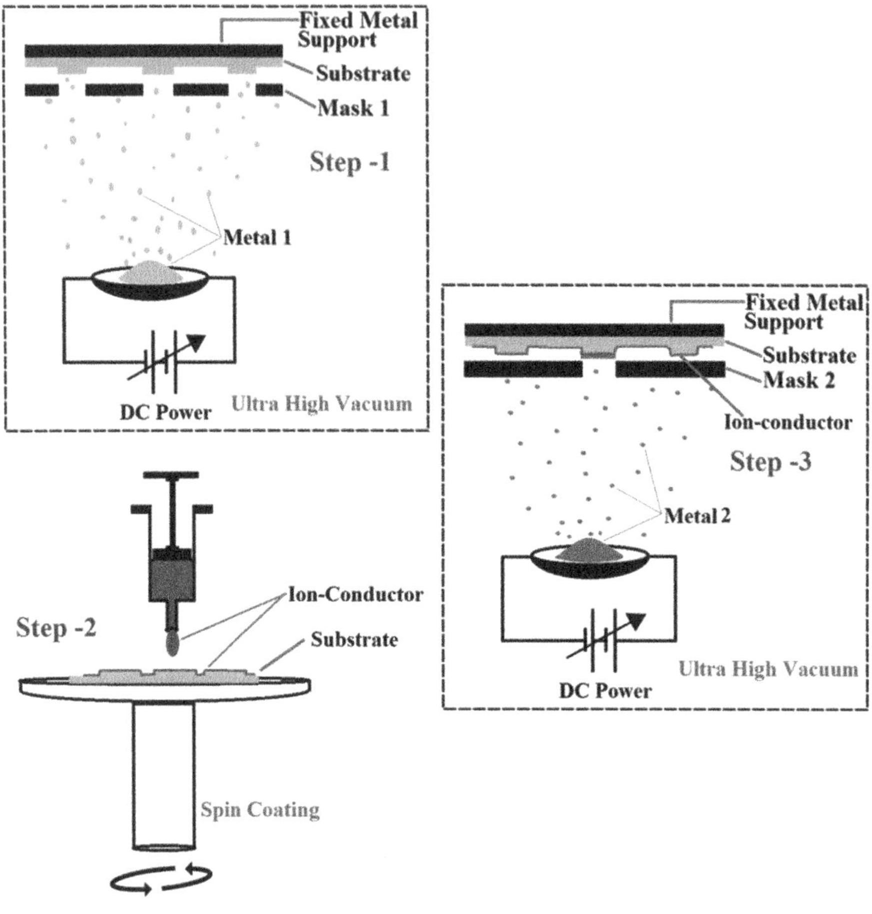

Figure 2.26. *Three-step process used for realization of CBRAM-based RF switches. For a color version of this figure, see www.iste.co.uk/jayakrishnan/technology.zip*

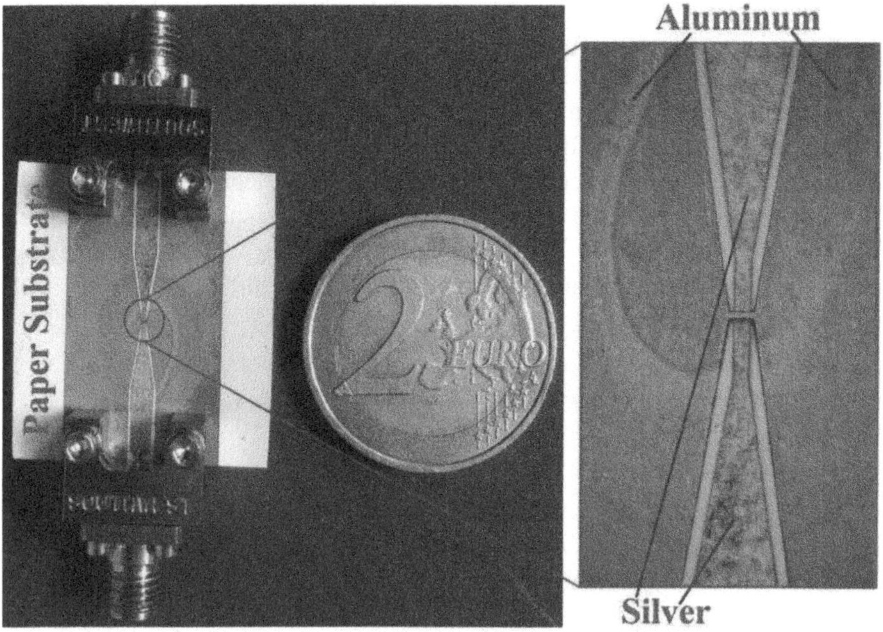

Figure 2.27. *CPW shunt mode RF switch on paper substrate with microphotograph of MIM switch area in the inset. Please note that the position of silver and aluminum layers are interchangeable (one on top of the other sandwiching the nafion layer), provided the switching pulses are applied with correct polarity. For a color version of this figure, see www.iste.co.uk/jayakrishnan/technology.zip*

2.6.2. *Results and discussion*

The RF response of the CBRAM-based CPW shunt mode RF switch on paper substrate is recorded with an Agilent ENA E5061B (3 GHz) Network Analyzer using a Short-Open-Load-Thru (SOLT) calibration scheme. The input power level used for the experiment is –15 dBm. Recorded S-parameter of the switch is depicted in Figure 2.28.

This switch shows an acceptable performance with insertion loss less than 3 dB in RF ON state and isolation greater than 15 dB up to 3 GHz in RF OFF state. One should note that this RF response is inclusive of the effects of connectors and feed lines. Here, the reader should not confuse the RF ON and RF OFF states represented in Figure 2.28 with the set (ON) and reset (OFF) states of the integrated CBRAM switch, which is in shunt mode in this design. RF ON corresponds to the reset (OFF) state of the integrated CBRAM switch, and vice versa in the case of the RF OFF state.

Filament resistance values of integrated CBRAM switches obtained through experimentation for each respective set (ON) and reset (OFF) state cycle are given in the brackets of the corresponding legends in Figure 2.28. Filament resistance values around 4 Ω and 40 kΩ, respectively, for set and reset states are observed in most cases. In this experiment, the highest observed set state resistance value is 10 Ω and the lowest reset state resistance value is 40 kΩ.

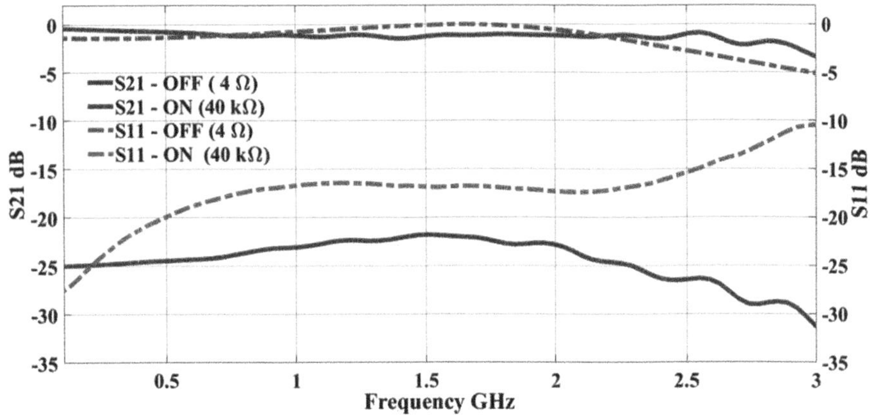

Figure 2.28. *Experimentally observed S-parameters of a CBRAM-based CPW shunt mode RF switch on paper substrate: S21 and S11. Filament resistance of each state is given in brackets. For a color version of this figure, see www.iste.co.uk/jayakrishnan/technology.zip*

A better qualitative analysis of this realization is tabulated in the form of a comparison with the realization of the same design on an FR4 substrate, presented in section 2.2, and with the state of the art (Nessel and Lee 2011) in Table 2.3. We can see in Table 2.3 that the presented realization of the CBRAM-based CPW shunt mode RF switch on flexible paper substrate has acceptable performance in comparison to the realization of the same design on FR-4 substrate.

Features	Realization on paper substrate	Realization on FR4 substrate	CBRAM state of the Art RF switch (Nanoionic Switch) (Nessel and Lee 2011)
Set state resistance (average)	7 Ω	2–5 Ω	10 Ω
Actuation voltage	~10 V/–20 V*	~10 V/–20 V*	~1 V
ON state RF insertion loss (average)	1.2 dB	0.5 dB	0.5 dB
OFF state RF isolation (average)	–23 dB	−16 to −34 dB	–35 dB
Operating frequency range	DC to 3 GHz	DC to 3 GHz	DC to 6 GHz
Minimum fabrication size (resolution) requirement	100 μm	100 μm	10 μm
Alignment accuracy required for fabrication	Low	Low	Very high
Number of steps in fabrication process and fabrication environment	3 (Ambient room)	6 (Ambient room)	More than 8 (Clean room process)
Power consumption to change state	μW/mW*	μW/mW*	μW
Power consumption to maintain state	0	0	0
Fabrication cost (laboratory perspective)	Low	Low	High
Manufacturing cost (industrial perspective)	Low	Low	Moderate
*For set and reset states, respectively.			

Table 2.3. *Comparison of fabricated CBRAM-based CPW shunt mode RF switches on paper substrate to the state of the art, and realization on rigid FR-4 laminates*

In addition, through the modified fabrication technique used for this realization, the number of process steps is reduced to half in comparison to the former process used for realization on FR-4 laminates. Minor degradation observed in the RF ON state insertion loss can be justified by the losses inherent to a paper substrate, and the fact that this design geometry is originally optimized for 50 Ω characteristic impedance on FR4 substrates.

Nevertheless, the outcome of this realization experiment for CBRAM-based CPW shunt mode RF switches on paper substrate displays affirmative outcomes, which could promise a magnificent future for realization of similar fully passive and non-volatile electronically reconfigurable RF and microwave devices for practical applications at low cost.

2.7. Application example: design and realization of solid-state non-volatile SPDT switch

An application example of the developed CPW shunt mode RF switch is demonstrated in this section. A well-known –3 dB Wilkinson power divider-based SPDT switch topology integrated with CBRAM/MIM-based shunt mode switches is presented herewith.

Here, we also present an insight on how the equivalent electrical model presented in section 2.3 could be used for low computational-effort simulations of CBRAM-based RF devices. Figure 2.29 shows the topology of a proposed SPDT switch based on a Wilkinson power divider optimized for operation around 300–800 MHz. The design is initially simulated (full wave) in CST Microwave Studio. The electrical equivalent circuit given in Figure 2.17 is used to model a CBRAM/MIM switch as shown in Figure 2.29. The switch part that includes the 600-nm thick nafion layer and 100 μm wide shunt line is removed, retaining tapering characteristics of the transmission line and ground plane. The MIM cell is then integrated in the form of a lumped RC parallel network with R and C values equivalent to the values reported in section 2.3.

Such a model saves a lot of the time required to mesh and compute the very thin 600 nm layer of nafion and 100 μm wide shunt line and nanometer-sized filament formations. A full structure simulation including all layers may be undertaken to map exactly the characteristics of the device, after a desired performance is optimized using this model. Nevertheless, we can observe that the simulation results using this technique are in agreement with the experimentally obtained results of the SPDT switch, as described below.

The fabrication of the proposed CBRAM-based electronically reconfigurable SPDT switch was accomplished using the process steps described in section 2.2.2 for rigid substrates, on a general purpose 1.6-mm thick FR-4 substrate with copper cladding on both sides. A photograph of the realized device is shown in Figure 2.30.

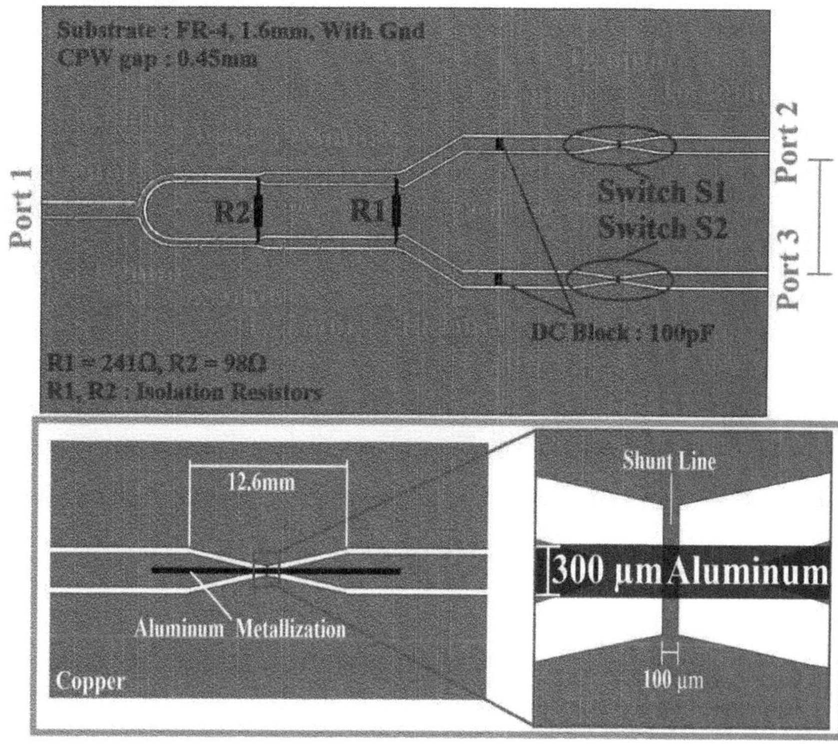

Figure 2.29. *Topology of CBRAM/MIM-based SPDT switch. Inset in green border shows the CPW shunt mode RF switch topology integrated as S1 and S2 in the main design. For a color version of this figure, see www.iste.co.uk/jayakrishnan/technology.zip*

DC pulses for operation of the integrated CBRAM switch are fed to the device directly through SMA connectors, from the Keithley 2400 source meter as explained in section 2.2. Two 100 pF DC blocking capacitors are integrated into the transmission line to avoid potential short circuiting of set and reset pulses among the divider transmission line, which does not have an ohmic isolation among the connecting ports at low frequencies. Isolation resistors for the Wilkinson power divider (Pozar 2005) are added using SMD components of appropriate values. The RF response of realized SPDT switch is recorded using the Agilent ENA E5061B (3 GHz) Network Analyzer using SOLT calibration scheme at a power level of –15 dBm. Recorded transmission and isolation characteristics of the SPDT switch along with full-wave electromagnetic simulation results are depicted in Figures 2.31 and 2.32, respectively.

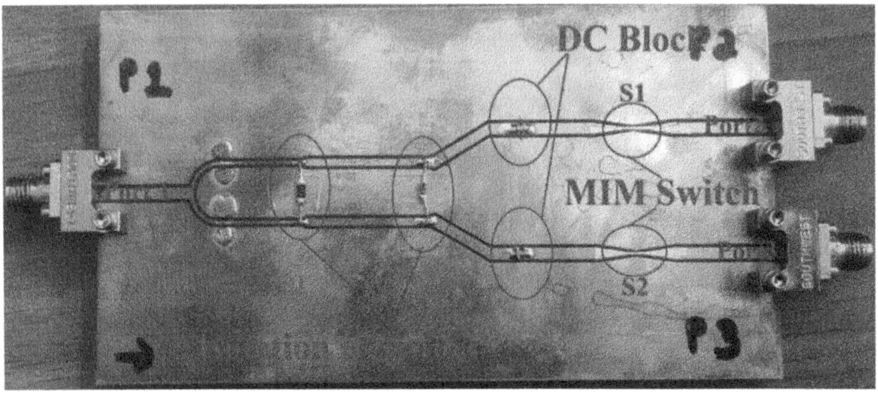

Figure 2.30. *Photograph of fabricated CBRAM-based electronically reconfigurable non-volatile SPDT switch. For a color version of this figure, see www.iste.co.uk/jayakrishnan/technology.zip*

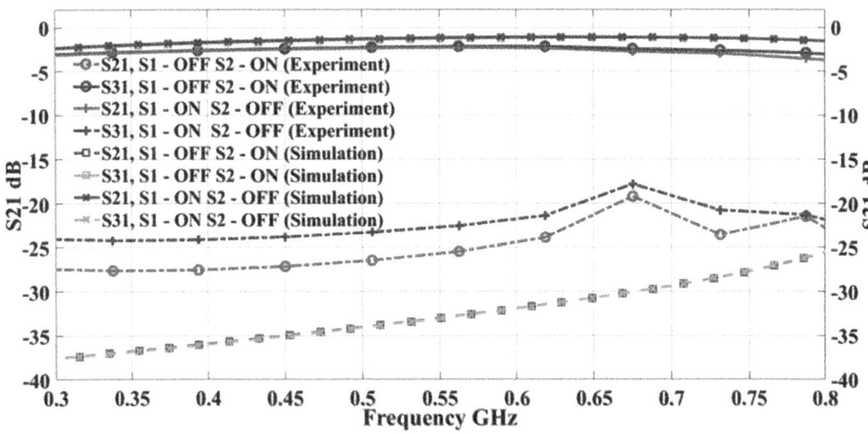

Figure 2.31. *RF transmission characteristics of fabricated CBRAM-based non-volatile SPDT switch. For a color version of this figure, see www.iste.co.uk/jayakrishnan/technology.zip*

From Figure 2.31, we can observe that the device exhibits 3 dB power splitting ratio and demonstrates SPDT switching characteristics for a bandwidth of about 500 MHz (90.9%) centered at 550 MHz (from 300 MHz till 800 MHz). The SPDT shows an average ON state transmission loss around 3 dB and an OFF state isolation greater than 17 dB in the entire bandwidth. Minor degradation in the performance is observed due to associated fabrication tolerances. The output port isolation (S23 and S32) is greater than 25 dB in all states of the SPDT switch as shown in Figure 2.32. This experiment affirmatively proves the concept of a solid-state SPDT switch using CBRAM/MIM switches, and this idea could be further extended to the development of single pole multiple throw (SPMT) switches, multiplexers and similar devices using the simplified fabrication techniques explained in this chapter.

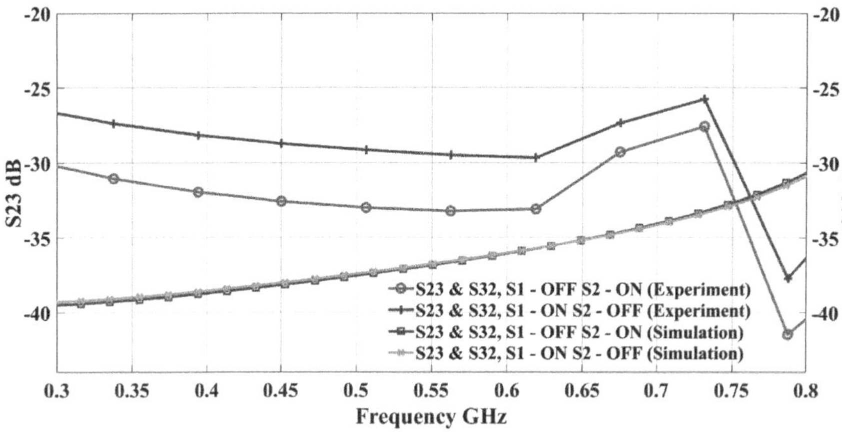

Figure 2.32. *Isolation characteristics of fabricated CBRAM based non-volatile SPDT switch. For a color version of this figure, see www.iste.co.uk/jayakrishnan/technology.zip*

2.8. Conclusion

In this chapter, we have successfully presented the real-world implementation challenges of a low-cost non-volatile RF switch and the techniques defined to overcome them. A simple technique for fast prototyping of CBRAM/MIM-based solid-state RF switches on classic–rigid substrates such as FR-4 is presented. Then a technique compatible with industrial processes for mass production is demonstrated for manufacture of RF switches on most substrates including flexible laminates like paper. We discuss the electrical equivalent modeling of MIM cells and the idea of using the same in electromagnetic simulations for fast and low computational effort simulations. An application of the proposed switch to realize a solid-state non-volatile SPDT switch is also shown affirmatively. These results affirm the feasibility of using CBRAM/MIM solid-state non-volatile switches for low-cost passive RF applications. The presented techniques for realization are compatible with standard industrial processes and do not require a clean room environment in any stage of production. Presented techniques could be easily adapted for high-speed roll-to-roll process for mass production of similar devices. In any case, although not as a mandatory requirement, a "clean room" support can be adopted for the optimization of standard fabrication steps to avoid the uncertainties related to the accuracy limit of general industrial processes and for quality control at desired production stages.

As a concluding remark of this chapter, we would like to point out the fact that the time in which these CBRAM RF switches could be directly printed along with passive microwave devices such as reconfigurable chipless RFID tags and reconfigurable antennas is not very far in the future. All the materials used for the layer-by-layer construction of the switch are potentially printable using an inkjet printer or similar techniques. The MIM layers could be realized by printing with metallic inks and ion conductor solutions.

In the following chapters, we present applications of the CBRAM/MIM switch using the proposed fabrication techniques in electronically rewritable/ reconfigurable passive microwave devices, including electronically rewritable chipless RFID labels, reconfigurable filters and pattern steerable antennas.

3

Solid-State Rewritable Chipless RFID Tags: Electronically Rewritable RF Barcodes

3.1. Introduction: chipless RFID technology

Radio frequency identification (RFID) is a technology that has been around for several decades. Inspired by radio detection and ranging (RADAR) techniques, RFID is the later successor of Identification Friend or Foe (IFF) technology used during Word War II. RFID is a very mature and diverse field in its current state (Finkenzeller 2010; Perret 2014). In general, the RFID scheme consists of an RF identifier in the form of a tag. The tag can either be active or passive. In the former case, the tag generates a specific modulated digital sequence that would be identified by a dedicated reader equipped with an antenna. In the latter case, the tag is just a scatterer, without any onboard power source, which would re-modulate and backscatter an interrogation energy sequence originating from a reader antenna.

In this study, we focus more on the passive RFID tag technology. The passive tags can be further categorized as chip-based and chipless tags. In the first type, the tag is composed of an identifier Integrated Circuit (IC) Circuit, which is generally a high-level semiconductor chip, which is impedance matched to an antenna. The reader illuminates the tag with an RF energy sequence (say a plane wave). The IC on the tag, upon the reception of this energy via antenna, scavenges a small part of it for power supply needs. The rest of the energy is remodulated with a specific identifier/code and retransmitted to the interrogator (say in the form of an amplitude shift keying or similar), which would now work in the reader mode and generate appropriate action according to the received rescattered RF energy (Finkenzeller 2010; Karmaker 2016).

In the case of chipless RFID tags, the identifiers are in the form of specific metallic or dielectric stacks or patterns that exhibit narrow band resonance, or group delay or phase encryption at certain predefined frequencies in a band of frequency (electromagnetic signature) and subdivided to certain windows to represent a bit pattern (Perret 2014; Karmaker 2016; Vena *et al.* 2016). This tag is also illuminated in similar way as the chipped tag, but with a wide-band RF signal in most cases, and the backscatter is received and interpreted using time domain or frequency domain techniques to identify the tag. Chipless RFID tags could be further categorized into surface acoustic wave (SAW)-based tags (Plessky and Reindl 2010), tags using filter network or delay lines connected to antennas (Preradovic *et al.* 2009; Nijas *et al.* 2012; Nair *et al.* 2013), multiscatterer-based tags (Perret 2014; Vena *et al.* 2016) and so on. Here, we emphasize multiscatterer-based chipless RFID tags, which are composed of narrow band metallic resonator structures, which at the same time function as the antenna and encoding particles (Vena *et al.* 2016; Islam and Karmakar 2012; Islam and Karmakar 2015). These tags are also known as RF-encoding particle (REP)-based tags, and are hence synonymously referred to as multiscatterer-based or multiresonator-based tags in this book.

Multiscatterer-based chipless RFID tags could also be called RF barcodes (Jalaly and Robertson 2005; Preradovic and Karmakar 2010) due to their similarity in appearance and operation to optical bar codes, except for the operating frequency/wavelength and information coding technology. However, chipless RFID technology could be considered to be a leap forward compared to the classical optical bar code technology, including added functionalities like the ability to integrate sensors, electronic reconfigurability/rewriteablity, non-line-of-sight readability and so on (Jalaly and Robertson 2005; Vena *et al.* 2011, 2012b; Tedjini *et al.* 2013; Amin *et al.* 2014a; Karmaker 2016; Borgese *et al.* 2017; Alves *et al.* 2018; Jayakrishnan *et al.* 2018a, 2018b).

Our study is specifically focused on the electronic re-configurability of chipless RFID tags, with integrated solid-state conductive bridging random access memory/metal insulator metal (CBRAM/MIM) switches

(Jayakrishnan *et al.* 2017; Vena *et al.* 2012a; Perret *et al.* 2015). We aim toward the realization of "fully printed solid-state and passive electronically rewritable chipless RFID tags", at an acceptable cost and size, on rigid as well as flexible substrates. The approaches we have adopted and the results obtained are explained in detail in this chapter.

3.2. Chipless RFID reader system used in this experiment

The simple form of RF equipment arrangement, similar to a bistatic RADAR setup, used in this study to read a chipless RFID tag is depicted in Figure 3.1. This system consists of two wideband horn antennas connected to a Network Analyzer. The antennas are co- or cross-polarized depending on the characteristics of device under test (tag of interest). A predefined wideband frequency sweep is illuminated on the tag using the transmitter antenna and the backscattered power amplitude and phase information is captured through the receiver antenna in the form of two-port scattering parameters. From this information, the true RCS (radar cross-section) of the tag under test could be extracted using equation [3.1] (Vena *et al.* 2011).

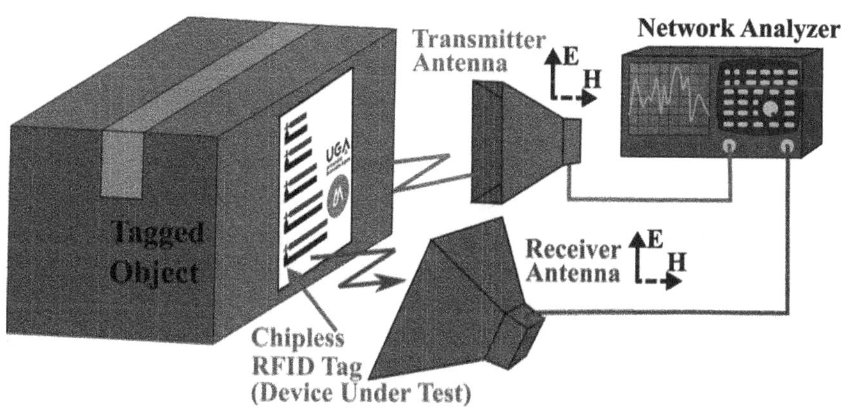

Figure 3.1. *Block diagram of RF equipment setup similar to bistatic radar used for RCS measurement of chipless RFID tags. For a color version of this figure, see www.iste.co.uk/jayakrishnan/technology.zip*

The RCS of a chipless RFID tag is an efficient parameter in determining the readability of a given tag in terms of read range for a given power using the well-known radar equation (Balanis 2005). The RCS of a target only depends on the target characteristics and not on the reader. Therefore, by considering the RCS as the identity of a tag, we have the convenience of it being reader independent. Any typical RCS measurement system with a desired accuracy could be used to read such a chipless RFID tag.

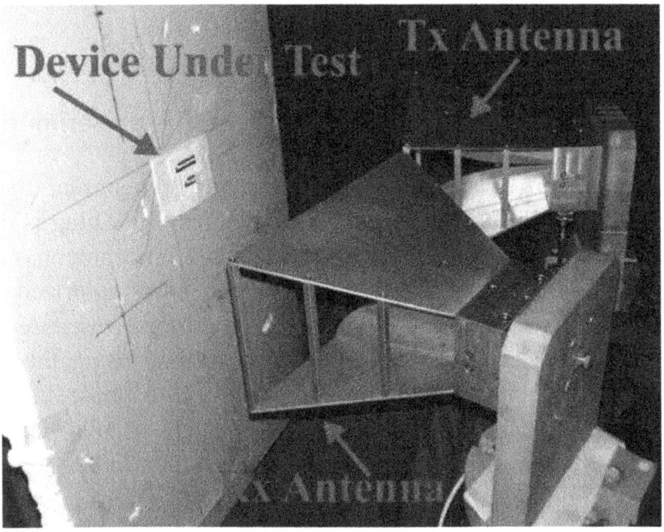

Figure 3.2. *Photograph of bistatic radar setup used for RCS measurement of chipless RFID tags. The Tx (transmitter) and Rx (receiver) antennas are connected to a 2-port network analyzer, which is not seen in this picture. For a color version of this figure, see www.iste.co.uk/jayakrishnan/technology.zip*

We use this measurement setup owing to its simplicity in setup and accuracy of results, with the equipment readily available in our laboratory. A photograph of the setup used for experiment is shown in Figure 3.2. This setup could be configured without discrimination in a real-world environment or in an anechoic chamber, with respect to the desired experiment.

$$\sigma^{tag} = \left[\frac{S_{21}^{tag} - S_{21}^{isolation}}{S_{21}^{ref} - S_{21}^{isolation}}\right]^2 \cdot \sigma^{ref} \qquad [3.1]$$

In [3.1], σ^{tag} is the complex RCS value of the tag under test, and σ^{ref} is the complex RCS value of a reference metallic plate of known dimensions calculated either analytically or through full-wave electromagnetic simulation studies. S_{21}^{tag} and S_{21}^{ref} are the complex S_{21} values captured, respectively, with the tag under test and reference plate in "device under test" position as shown in Figure 3.1, and $S_{21}^{isolation}$ is the S_{21} value captured from the setup without any devices under test, but with all other features undisturbed.

In addition to such simple reader setup, some efficient and low-cost designs of stand-alone chipless RFID readers are reported recently from our group as in Garbati *et al.* (2019), which could be fabricated as a handheld or desktop model upon demand. An example of such a reader is shown in Figure 3.3. Apart from investigations of tag and reader technologies, advance postprocessing of backscatter data from chipless RFID tags for accurate detection among a noisy environment is also a hot topic of recent research (Ramos *et al.* 2016).

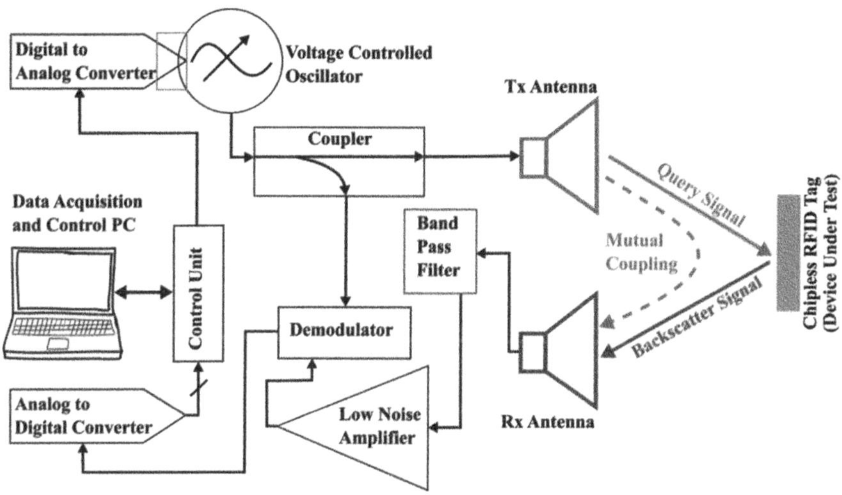

Figure 3.3. *Block diagram of a modern compact chipless RFID reader (redrawn from Garbati* et al. *2019)*

3.3. Realization of solid-state electronically rewritable chipless RFID tags

In this section, we present the design and realization of an electronically reconfigurable or rewritable REP for chipless RFID applications. We present a rewritable resonator inspired by the well-known "C" resonator (Vena *et al.* 2011). The electronic reconfigurability/rewriteablity of chipless RFID is a crucial factor that distinguishes this technology (often referred to as RF barcodes) from the classic optical barcodes and opens the way for versatile new applications. The possibility of electronic reconfiguration helps in reducing the cost per unit tag through bulk manufacture, and makes it possible to reuse a tag for several applications. An entire batch of chipless RFID tags could be manufactured as a single copy, and it could be written or rewritten with the desired code at the time of application by the user. This is in place of the individual attention required to mechanically reconfigure tags, utilizing conductive printing, etc., at a higher cost.

3.3.1. *Electronically rewritable chipless RFID tags on classic rigid substrates*

A one bit solid-state electronically rewritable chipless RFID tag on FR-4 substrate is presented in this section. The geometry of two versions of the tag with one REP each (with different frequencies of operation) is depicted in Figure 3.4, namely Tags 1 and 2. Measurements of these tags are summarized in Table 3.1, with respect to the geometry in Figure 3.4.

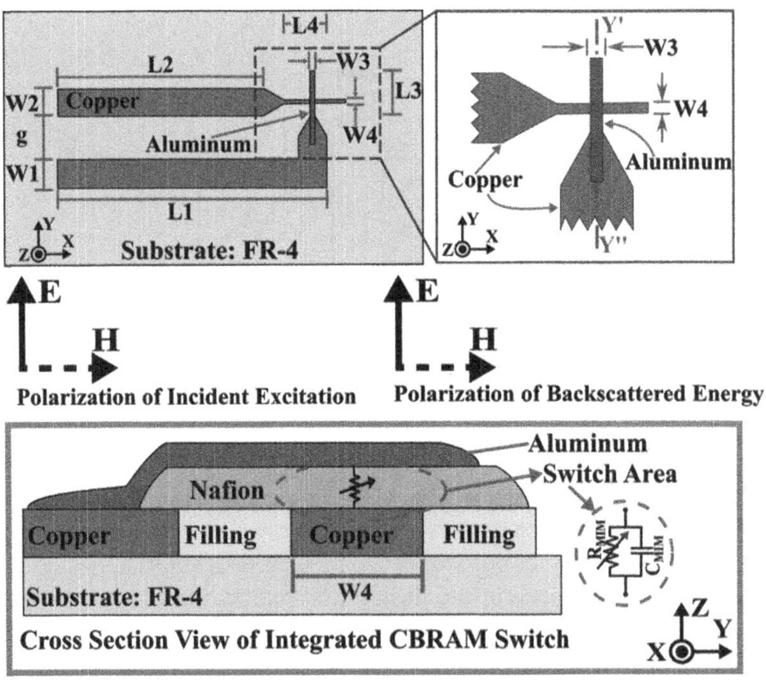

Figure 3.4. *Geometry of electronically rewritable resonator for chipless RFID tag. Inset in green border shows the cross-section (along the plane Y' – Y'') of how the aluminum layer is attached to copper and the resulting formation of integrated CBRAM switch. For a color version of this figure, see www.iste.co.uk/jayakrishnan/technology.zip*

The REP resonates similarly to a quarter wavelength rectangular slot (on a metallic plane) or a half wavelength shorted (and folded) dipole, when illuminated by an electromagnetic plane wave polarized along the Y axis as shown in Figure 3.4. A CBRAM/MIM cell is integrated into the center of this metallic resonator, to tune its electrical length depending on switch state, i.e. set (ON) or reset (OFF). The MIM switch is integrated into the resonator by tapering trace width to reduce the effective capacitance due to overlapping sections. The switch is formed with copper as the active electrode (marked as width "w4") and aluminum as the inert electrode (marked as width "w3"), with a 600 nm layer of nafion sandwiched between them as the ion conductor layer, as shown in Figure 3.4. In this way, one could electronically reconfigure the resonator to resonate at two different frequencies. This tag with one resonator could resonate at two different frequencies representing one bit of information.

Parameters (mm) →	W1	W2	W3	W4	L1	L2	L3	L4	G
Tag 1	3	3	0.4	0.16	21.5	14.5	1.8	4	3
Tag 2	3	3	0.4	0.16	37	30	1.8	4	3

Table 3.1. *Measurement of solid-state rewritable chipless RFID tags on FR-4 substrate, as shown in Figure 3.5 (refer Figure 3.4 for geometry)*

One should note here that this resonator with an integrated MIM switch shows two different resonant frequencies, upon operation of this switch, for set (ON) and reset (OFF) states. This is due to the non-ideal behavior of the integrated MIM switch. These switches, unlike ideal switches, exhibit a considerable OFF (reset) state capacitance, which tunes the electrical length of the resonator, forcing it to resonate at a higher frequency with respect to the initial frequency. If an ideal switch with zero (or negligible) OFF state capacitance is used here, it would simply suppress the resonance of a particular resonator in its OFF state, or at least shift the resonance to very high (double) frequencies from the bandwidth of interest. This is due to the dissection of resonator length in the OFF sate of an ideal switch. The effect of the OFF (reset) state capacitance in the mechanism of the resonator is explained in detail in the electrical model analysis in the following section.

The design is optimized in full-wave electromagnetic simulation using CST Microwave Studio. A lumped resistor inserted into the copper-nafion-aluminum sandwich structure with values similar to the measured filament resistance of MIM switch (0.5 Ω and 30 MΩ) is used to operate set (low impedance) and reset (high impedance) states of the switch (similar to as explained in Chapter 2 for MIM switch simulations). An electromagnetic plane wave with electric field polarized as shown in Figure 3.4 is used as the excitation, and an RCS probe placed at a distance of 100 mm from the center of the resonator structure is used to capture the backscatter response.

Two resonators with different frequency of operation are realized for validation of the concept. These one bit tags are fabricated on classic FR-4 substrates in an ambient environment. The fabrication process is the same as optimized for realization of CBRAM/MIM RF switch as explained in Chapter 2 (see also Jayakrishnan *et al.* 2017). Photographs of realized tags with a zoomed microphotograph of the MIM switch area in the inset are shown in Figure 3.5.

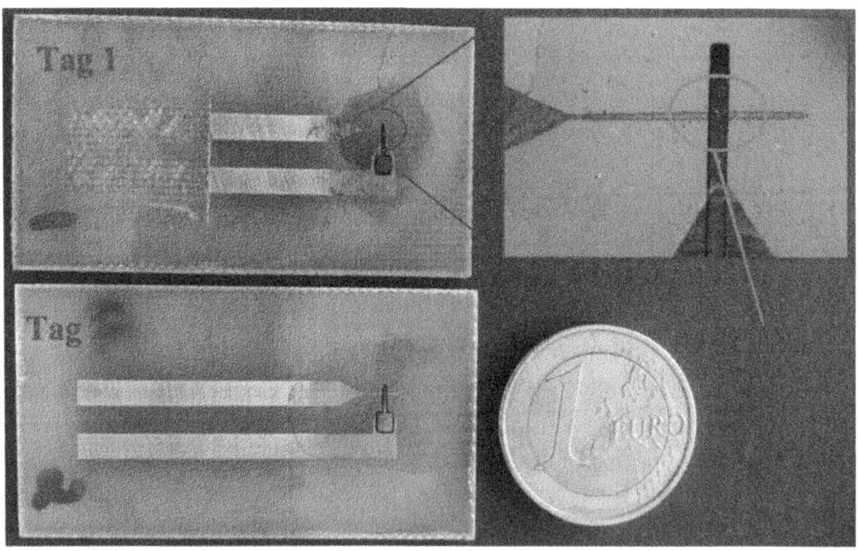

Figure 3.5. *Photograph of fabricated electronically rewritable chipless RFID tags on FR-4 substrates (inset shows microphotograph of MIM switch area). For a color version of this figure, see www.iste.co.uk/jayakrishnan/technology.zip*

The integrated MIM switch in fabricated rewritable resonators is set (to low impedance state) using a DC voltage ramp of +12 V with a compliance current limit of 0.5 mA and reset (to high impedance state) using a square pulse of –20 V with a compliance current limit of 0.15 A (Jayakrishnan *et al.* 2018a), as explained in Chapter 2. The pulse polarities are stated and applied with respect to the active electrode, among the active and inert electrodes. Switching pulses are applied physically using ordinary crocodile clips attached to metallic parts of the resonator.

3.3.1.1. *Results and discussions*

RF responses of the fabricated tags were measured using Agilent N5222A Network Analyzer, inside an anechoic chamber, in a bi-static radar setup (Vena *et al.* 2011), as shown in Figure 3.2, using two wideband horn antennas in co-polarization. The RCS backscatter of the tags obtained in the experiment along with the results of the simulation are shown in Figures 3.6 and 3.7 for Tags 1 and 2, respectively.

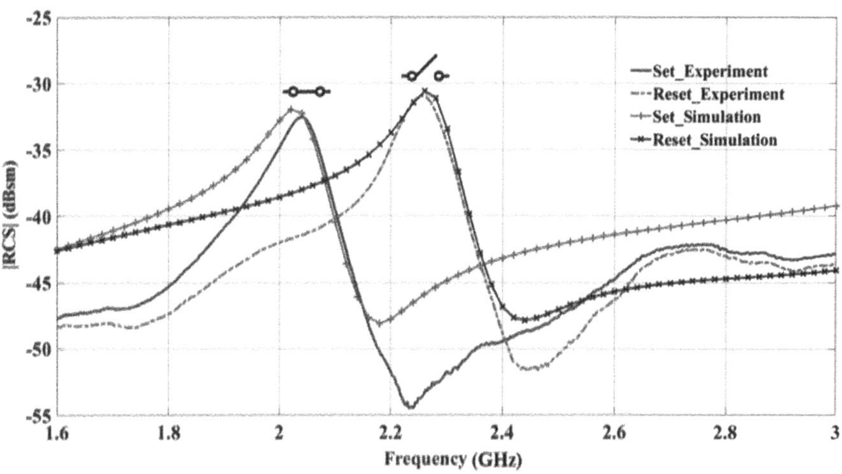

Figure 3.6. *Experimentally measured and simulated (full-wave) RCS response of electronically rewritable chipless RFID tag: Tag 1. For a color version of this figure, see www.iste.co.uk/jayakrishnan/technology.zip*

The magnitude of resonance frequency peaks in the experimentally obtained RCS backscatter is clear and high enough to be recognized with a tag reader in a real environment with read range of around 30 cm (Vena *et al.*

2011). Tags 1 and 2 show a displacement of, respectively, 200 and 140 MHz, among set (ON) and reset (OFF) state resonance frequencies. Experimental results are in good agreement with the results of the electromagnetic simulation. Each tag (with one resonator each) presented here represents one bit, and five such resonators could be integrated to a substrate the size of a standard commercial credit card, maintaining appropriate spacing among the REPs to avoid any mutual coupling issues. Moreover, each resonator could be further optimized to integrate more MIM switches to increase the number of bits per tag.

Here, both of the tags presented were switched for more than 50 cycles of operation and are observed to be still functional even after this number. Integrated MIM switches exhibit set (ON) state resistance among 1 and 10 Ω and reset (OFF) state resistance greater than a few MΩ in all cases. This is a considerably good number of repetitions for electronically rewritable chipless RFID tags for short-term identity or counter applications like in public transport tickets. This design was reported initially in Jayakrishnan *et al.* (2018a).

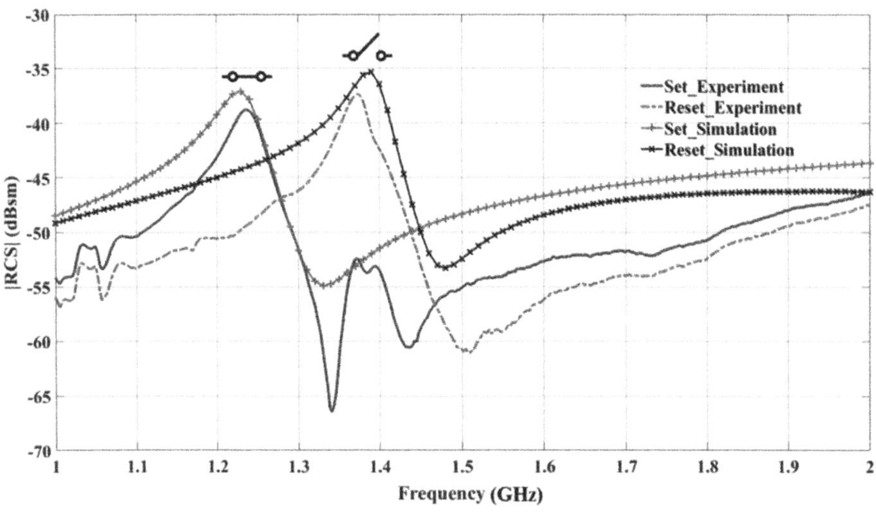

Figure 3.7. *Experimentally measured and simulated (full-wave) RCS response of electronically rewritable chipless RFID tag: Tag 2. For a color version of this figure, see www.iste.co.uk/jayakrishnan/technology.zip*

3.3.2. Electronically rewritable chipless RFID tags on flexible substrates

Next, we demonstrate the design and implementation of electronically rewritable chipless RFID tags on flexible and low-cost substrates such as paper, which is seen in our daily surroundings as a very common packaging material for goods, public transport tickets, visiting cards and so on. An RF identification tag using the proposed electronically rewritable REP could be easily fabricated on a paper substrate with adhesive background and could be attached to an object as a sticker, which forms a cheap and versatile electronically rewritable identifier.

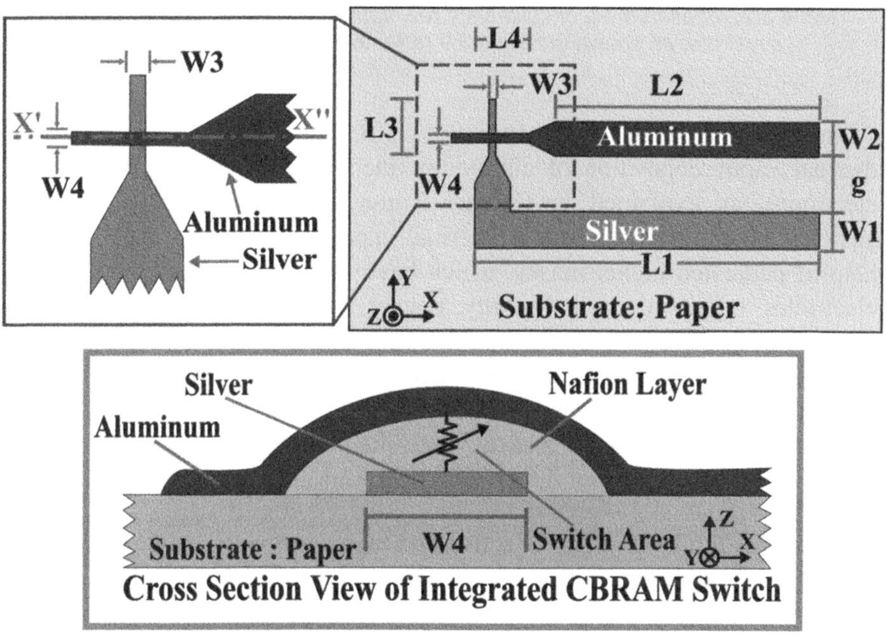

Figure 3.8. *Topology of electronically rewritable resonator for chipless RFID applications on paper substrate. Inset in green border shows the cross-section view along the plane X' – X", showing arrangement of metallic and ion conductor layers resulting in formation of integrated CBRAM switch. For a color version of this figure, see www.iste.co.uk/jayakrishnan/technology.zip*

The substrate used for realization of this rewritable chipless RFID tag is 80 g/m^2 (GSM) glossy paper. The geometry of resonators (REP) used for the realization is shown in Figure 3.8; this is similar to the design shown in

Figure 3.4 but with the difference that silver is used as the active electrode in place of copper. In this case, the tag consists of two resonators R1 and R2 whose length measurements are summarized in Table 3.2, and the tag could represent a total of four states or two bits of information.

Parameters (mm) →	W1	W2	W3	W4	L1	L2	L3	L4	G
R1	3	3	0.16	0.14	37	30	3	4	3
R2	3	3	0.16	0.14	18	11	3	4	3

Table 3.2. *Measurement of solid-state rewritable chipless RFID tags on paper substrate, as shown in Figure 3.9 (refer to Figure 3.8 for geometry)*

In this case, metal layers are grown on the paper substrate through thermal vapor deposition of aluminum (inert electrode) and silver (active electrode), as explained in Chapter 2 for realization of CBRAM/MIM switches on flexible substrates. Thermal vapor deposition is done with the help of dedicated nickel masks, which have an aperture of the topology of electrodes, engraved on it. Initially, one of the electrodes (either silver or aluminum) is deposited. After this, the ion conductor layer is made by spin coating of nafion resin solution (Sigma-Aldrich 2019) at a rate of 500 RPM for 30 s. The ion conductor layer is then air dried at 100°C for 2 min on a hot plate. After that, the opposite electrode (aluminum or silver) is formed by thermal vapor deposition of appropriate metal.

Note that the layer order of active and inert electrodes is interchangeable in the presented design, if the polarity of application of the switching pulse is retained properly with respect to appropriate electrodes. The thicknesses of deposited silver and aluminum layers are measured to be 1.3 μm and 1 μm, respectively, and that of deposited nafion is observed to be 600 nm. These measurements are carried out using a Dektak-150 mechanical profilometer. A photograph of a fabricated rewritable chipless RFID tag on paper is shown in Figure 3.9, with zoomed microphotograph of the MIM switch area in the inset.

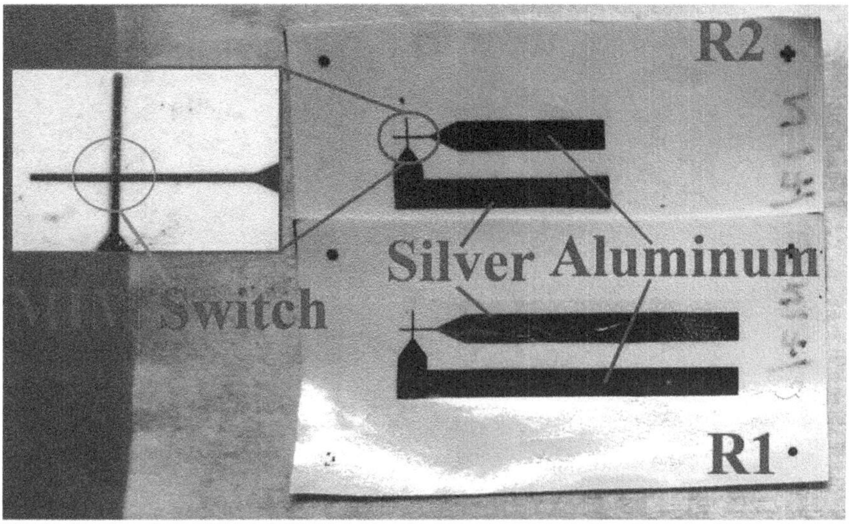

Figure 3.9. *Photograph of fabricated electronically rewritable chipless RFID tag on paper substrate, with microphotograph of MIM switch area in the inset*

3.3.2.1. *Results and discussions*

RCS response of fabricated tag is captured using the same setup as explained in section 3.2, and is depicted in Figure 3.10. One could observe that both set (ON) and reset (OFF) state resonance frequency peaks are clearly distinguishable for each resonator with adequate RCS magnitude level and are separated by at least 380 MHz for both switch impedance states. Filament resistance of MIM switch, obtained in experiment, for set (ON) state is observed to be 1–10 Ω, and 1–30 MΩ for reset (OFF) state, for several cycles. This rewritable chipless RFID tag represents four different combinations (two bits of information) depending on the impedance states of integrated MIM switch as shown in RCS plots. Results of full-wave electromagnetic simulation are shown in Figure 3.11 for comparison. We can see that there are slight differences in RCS level among the experimental and simulated response, and also with the realization of the same design on copper cladded FR-4 substrate explained in section 3.3.1. This could be justified with the following explanation.

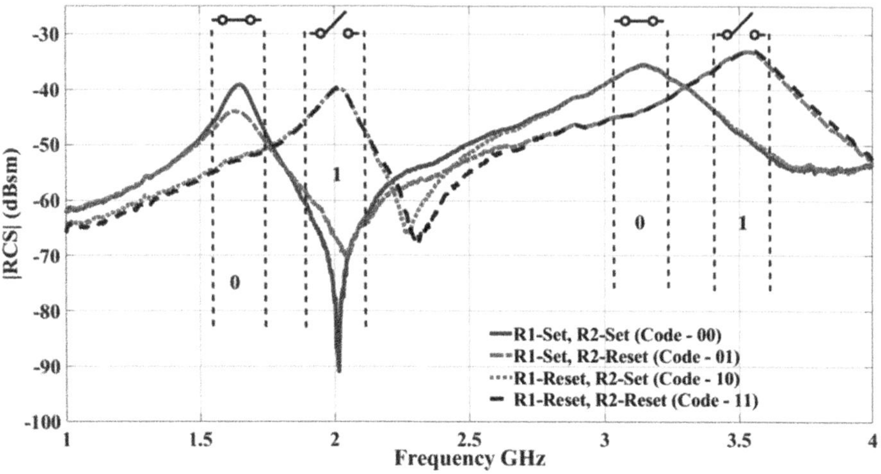

Figure 3.10. *Experimentally measured RCS response of electronically rewritable chipless RFID tag on paper substrate. For a color version of this figure, see www.iste.co.uk/jayakrishnan/technology.zip*

First, in this design we realize the metallic layers by thermal vapor deposition of silver and aluminum. For both these metal layers, their thickness is less than or at least at the limit of theoretical "skin depth" at the operating frequency. The thickness of silver and aluminum layers is, respectively, 1.3 and 1 µm in this realization, and the skin depth at 3 GHz for bulk silver and aluminum is 1.16 and 1.5 µm, respectively.

Second, the conductivity of thermal vapor deposited metals in this case is less than the bulk conductivity of original metals. This is due to variations in multiple control factors in the thermal vapor deposition process such as the surface roughness of substrate (paper in this case), defects among mutual attachment or linking of metal particles originating from respective vapor phase, contamination inside the thermal vapor deposition unit, adhesion between substrate and metal, purity of metals and so on. In our thermal deposition unit, the conductivity of measured samples on glass substrate is measured to be as follows: silver: 16e6 to 30e6 S/m; aluminum: 14e6 to 28e6 S/m. The naturally observed bulk conductivity of pure metallic silver and aluminum is, respectively, 63e6 and 37e6 S/m.

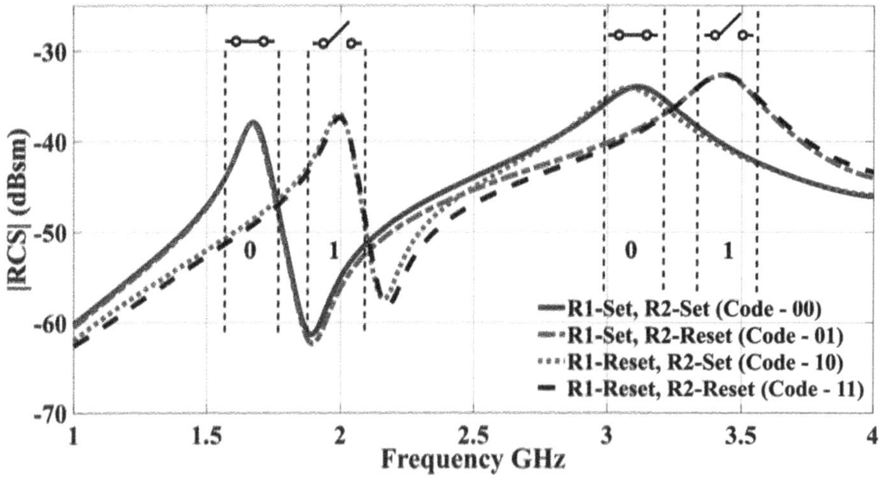

Figure 3.11. *Simulated (full-wave) RCS response of electronically rewritable chipless RFID tag on paper substrate. For a color version of this figure, see www.iste.co.uk/jayakrishnan/technology.zip*

The copper-nafion-aluminum-based tags explained above show a superior RCS level, as the thickness (35 μm) and conductivity (59e6 S/m) of the copper layer (which constitutes the majority of the tag surface) is much higher than what is used here, and the use of thermal deposited aluminum is limited only to a small part of the switch area in the former design. Moreover, for the simulated results of the chipless tag on paper substrate presented here, we also have the effect of the difference associated with the unaccounted losses with the paper substrate, used for the experiment, and the model of the paper substrate available with CST Microwave Studio used for simulation.

However, the rewritable chipless RFID tags on paper substrate presented here show a distinguishable RCS level, and affirmatively prove the feasibility of such an application. RCS levels of these tags could be improved further through optimizations such as using a paper substrate with better surface properties suitable for thermal vapor deposition, increasing thickness of deposited metals, choice of a resonator with better RCS and so on. These studies are in progress with the authors along with other future perspectives discussed in this book and are expected to be published soon.

Next, possible strategies for data encoding and advantages of using this concept in a commercial context through mass production, and the expected convenience, are explained in the following sections.

3.4. Effect of CBRAM/MIM filament resistance on RCS characteristics of presented electronically rewritable resonators

In this section, we discuss the effect of the MIM switch filament resistance on RCS amplitude characteristics of presented electronically rewritable resonators. This is an important aspect that can affect the read range and quality of detection of a rewritable chipless RFID tag. Figure 3.12 presents a plot of variation of RCS of a resonator, similar to R1 in section 3.3.2, as a function of filament resistance of an integrated CBRAM/MIM switch. This result is obtained in full-wave electromagnetic simulation with CST Microwave Studio by varying the filament resistance of the MIM switch.

Here, we can see that the REP is visible (i.e. shows a clear resonance frequency peak) in terms of RCS levels for a filament resistance value less than or equal to 10 Ω in the so-called set (low impedance) state and greater than a few kΩ in the reset (high impedance) state. In addition, from this observation, we can deduce that we could disable a particular REP of choice in a given chipless RFID tag using such resonators, by setting its filament resistance between 50 and 150 Ω. Moreover, we could use this strategy to introduce more coded states for a typical electronically rewritable REP utilizing this amplitude control of RCS magnitude, depending on the impedance control of corresponding integrated MIM switches.

As described in Chapter 2, at present we could perform such an impedance control of MIM switches through trial and error modulation of operating DC pulses. A reliable technique to predict the filament resistance of CBRAM/MIM switches is yet to be defined. With the present operating pulse design, emphasis is placed on achieving low resistance state less than 20 Ω in set (ON) state and greater than a few kΩ in reset (OFF) state, without damaging the switch. In addition to this, better techniques to achieve and predict a desired filament resistance for CBRAM/MIM switches are in progress, and are expected to be published in the near future.

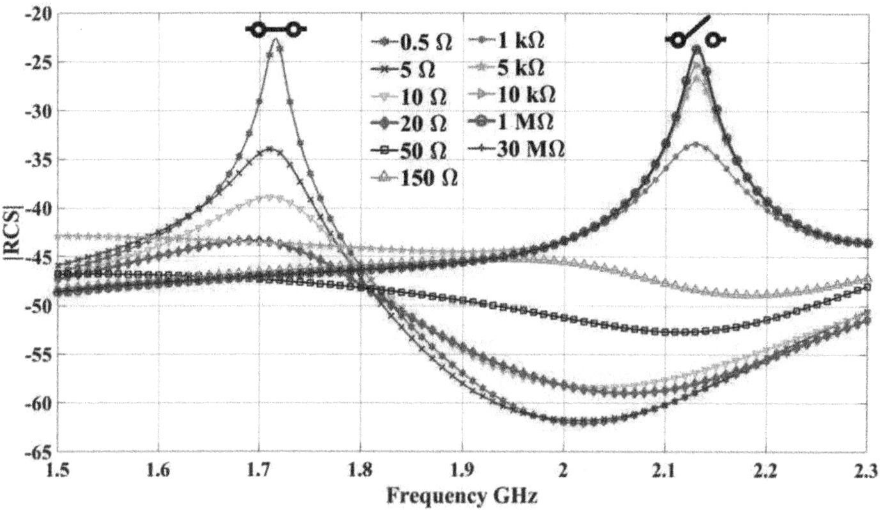

Figure 3.12. *Simulated RCS response of a single electronically rewritable resonator R1 showing amplitude variation as a function of filament resistance of an integrated CBRAM/MIM switch. For a color version of this figure, see www.iste.co.uk/jayakrishnan/technology.zip*

3.5. Electrical equivalent model of electronically rewritable chipless RFID tags

In this section, we present the electrical modeling of the proposed electronically rewritable chipless RFID tags mentioned in this chapter. In general, an electrical model of a typical chipless RFID tag composed of "N" number of C-shaped resonators (REPs) as shown in Figure 3.13, could be approximated as shown in Figure 3.16 (Vena *et al.* 2013b, 2016). This direct approximation of the principle resonance frequency of a "C" resonator-based tag is a combination of two time-separated resonant modes of REPs on the chipless RFID tags, namely structural mode and antenna mode. Separately resolved antenna and structural modes of a single resonator (REP) along with its combined electromagnetic response are illustrated in Figures 3.14 and 3.15, respectively, in terms of magnitude and phase response.

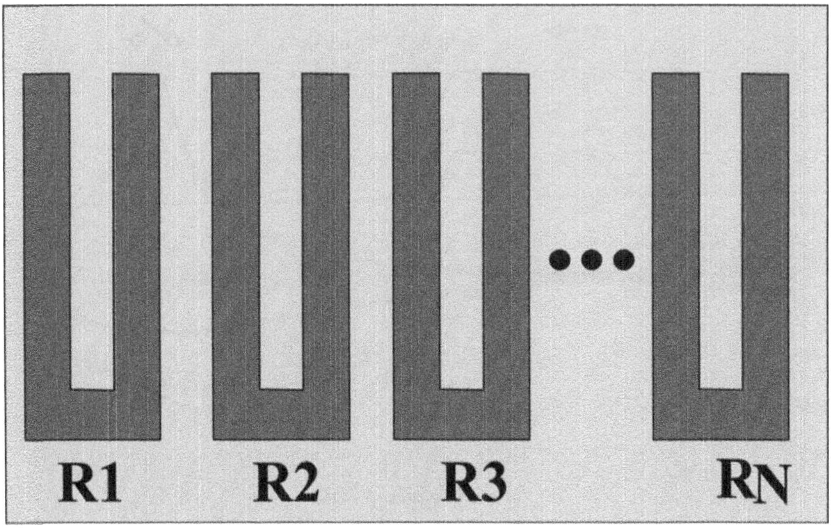

Figure 3.13. *Multiscatterer-based chipless RFID tag using "C"-shaped resonators*

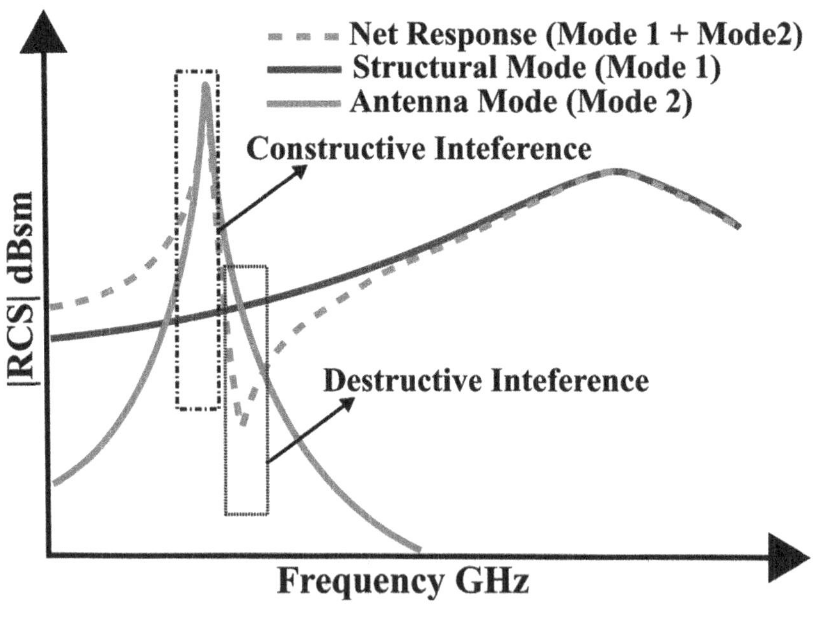

Figure 3.14. *Representation of resonance frequency of scattering resonator-based chipless RFID tag resolved into antenna mode and structural mode: magnitude plot. For a color version of this figure, see www.iste.co.uk/jayakrishnan/technology.zip*

The structural mode of the tag is a wideband resonant response, as a result of immediate scattering of incident electromagnetic waves from the apparent metallic surface, and is a phenomenon similar to optical reflection. On the other hand, the antenna mode is a narrowband resonant response of the tag, as a result of backscattering of the energy stored from incident waves in the resonators and is shifted in time with respect to the structural mode. In summary, the net RCS resonant response of a resonator would be an interference pattern of both these modes, with respect to their corresponding instantaneous phase angle, as shown in Figures 3.14 and 3.15.

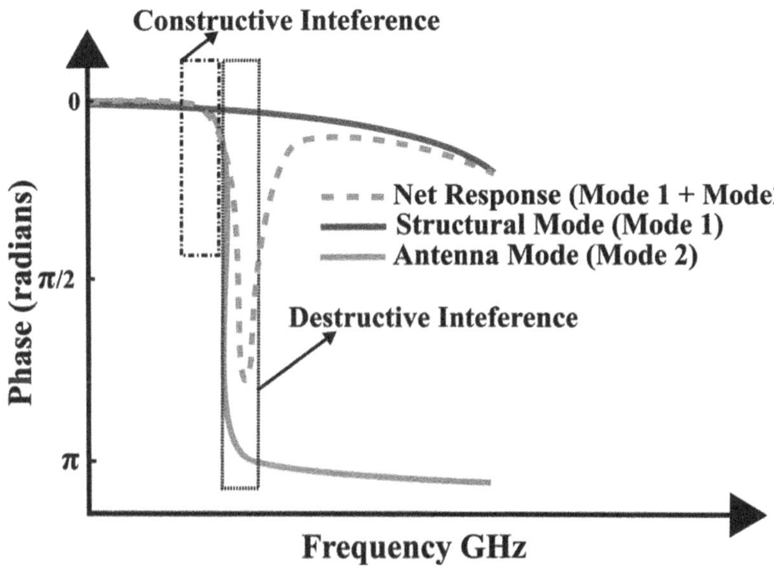

Figure 3.15. *Representation of resonance frequency of scattering resonator-based chipless RFID tag, resolved into antenna mode and structural mode: phase angle plot. For a color version of this figure, see www.iste.co.uk/jayakrishnan/technology.zip*

Based on this theory, the electrical model of a multiscatterer-based chipless RFID tag would thus consist of a model of a wideband structural mode resonant network for the whole tag, and cascaded narrow band antenna mode networks for each resonators present on the tag. Figure 3.16 shows such an approximation with an RLC equivalent for the structural mode of the entire tag and cascaded RLC equivalents for the antenna mode of each resonator present on the tag. The voltage sources connected to each RLC network model the fraction of energy backscattered with each mode. This model accurately demonstrates the principal resonance of the tag for each resonator, but does not take into account the higher order resonances which occur at two times the principal resonant frequency for "C"-shaped resonators. Nevertheless, this model is a good approximation to be used to model the proposed rewritable chipless RFID tags. It will be apparent in the following sections how this model could help one in predicting the resonance behavior of each resonator on a chipless RFID tag and in simplifying the data encoding process. For more details on the theory of RF response modeling of multiscatterer-based chipless RFID tags, the reader may refer to Vena *et al.* (2016).

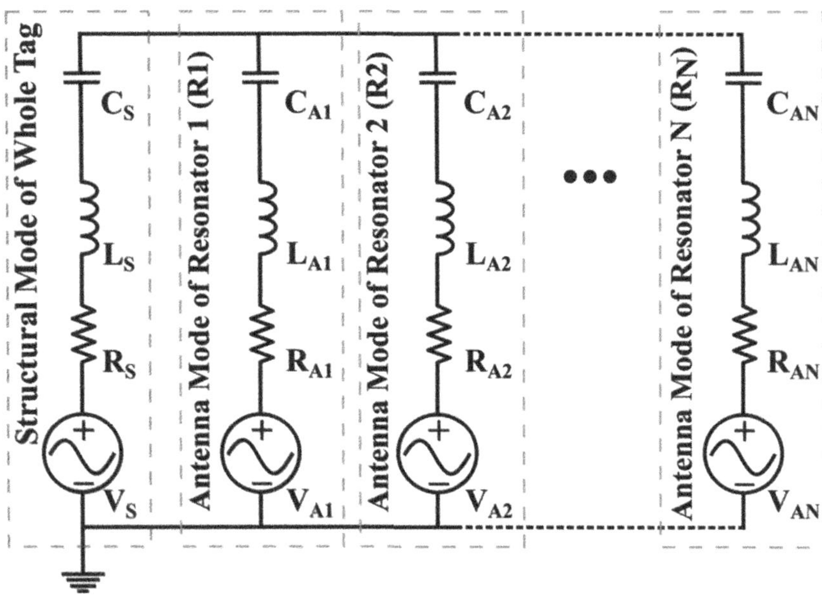

Figure 3.16. *Electrical equivalent model of "C"-shaped multiresonator-based chipless RFID tag shown in Figure 3.13*

The electrical equivalent model of the electronically rewritable chipless RFID tag with integrated CBRAM/MIM switches, presented in this book, could be approximated by inserting the equivalent circuit of an MIM cell (which is an RC parallel network formed by C_{MIM} and R_{MIM}) in the antenna mode of each resonator as shown in Figure 3.17. Here, R_{MIM} and C_{MIM} are, respectively, the filament resistance and equivalent capacitance due to the geometry of the integrated MIM cell as explained in Chapter 2.

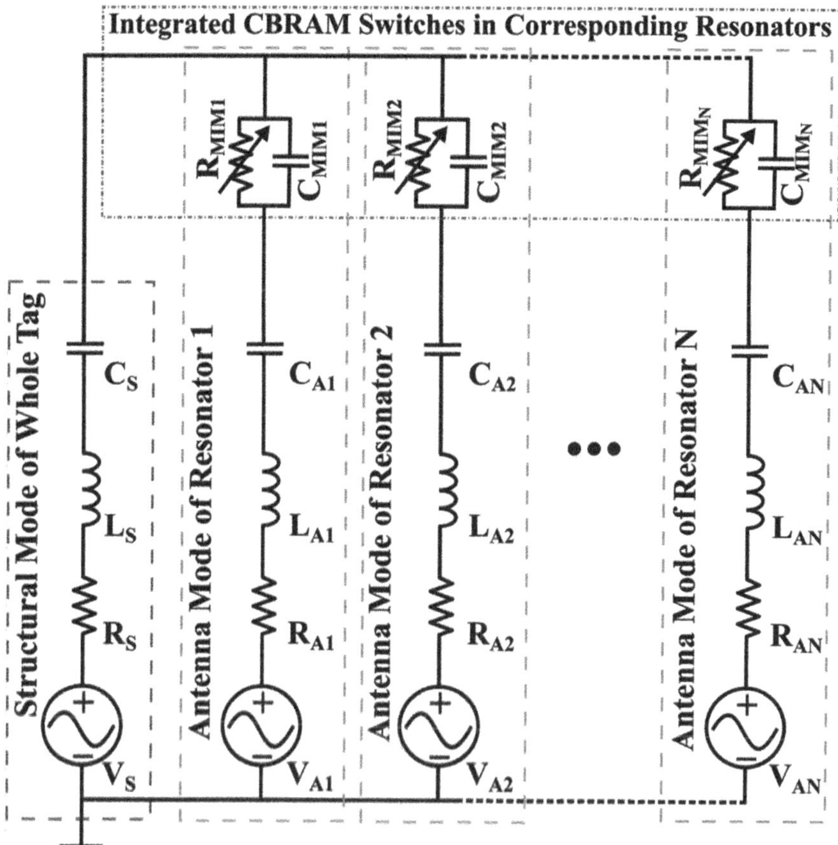

Figure 3.17. *Electrical equivalent model of electronically rewritable "C"-shaped resonator-based chipless RFID tag presented in this chapter*

In the evolution process of the presented model, the initial values of C_{MIM} were chosen to be similar to the parallel plate capacitance of the MIM cell structure, calculated from the geometry of electrodes and the dielectric constant of the ion conductor used as explained in Chapter 2. Starting values of R_{MIM} resistance were chosen similarly to the experimentally measured DC filament resistance across the MIM cell, respectively, for set (ON) and reset (OFF) states. Other lumped element values were chosen from general engineering calculation of RLC series resonances. All these component values were then eventually optimized for the best performances through fitting with experimentally obtained results and are depicted in Table 3.3 for the tags fabricated on FR-4 described in section 3.3.1 and in Table 3.4 for tags fabricated on paper substrate described in section 3.3.2. PSpice circuit simulator was used to develop and operate this model. Results of the simulation of the electrical equivalent model, normalized to experimental results, and also with experimental results for comparison, for electronically rewritable chipless RFID Tags 1 and 2 presented in section 3.3.1 (fabricated on FR-4 substrate), are given, respectively, in Figures 3.18 and 3.19. Figure 3.20 shows the results of simulation of the electrical equivalent model for electronically rewritable chipless RFID tags on paper substrates presented in section 3.3.2.

Parameters	Tag 1	Tag 2	Parameters	Tag 1	Tag 2
R_S	20 Ω	20 Ω	R_{A1}	5 Ω	8 Ω
L_S	1 nH	1 nH	L_{A1}	14 nH	35 nH
C_S	0.2 pF	0.2 pF	C_{A1}	0.43 pF	0.473 pF
V_S	30 mV, 0°	30 mV, 0°	R_{MIM1}	0.5 Ω/1 MΩ*	0.5 Ω/1 MΩ*
V_{A1}	1.5 mV, 0°	1.5 mV, 0°	C_{MIM1}	1.98 pF	1.98 pF
*For set (ON) and reset (OFF) states of CBRAM/MIM switch.					

Table 3.3. *Optimized electrical equivalent model component values of rewritable chipless RFID tag on FR-4 substrate, as described in section 3.3.1*

Parameters	Value	Parameters	Value	Parameters	Value
R_S	40 Ω	R_{A1}	8 Ω	R_{A2}	15 Ω
L_S	1 nH	L_{A1}	14 nH	L_{A2}	10 nH
C_S	0.2 pF	C_{A1}	0.66 pF	C_{A2}	0.255 pF
V_S	500 mV, 0°	R_{MIM1}	1 Ω/1 MΩ*	R_{MIM2}	1 Ω/1 MΩ*
V_{A1}	80 mV, 0°	C_{MIM1}	1.3 pF	C_{MIM2}	1 pF
V_{A2}	200 mV, 0°	-			
*For set (ON) and reset (OFF) states of CBRAM/MIM switch.					

Table 3.4. *Optimized electrical equivalent model component values of rewritable chipless RFID tag on paper substrate, as described in section 3.3.2*

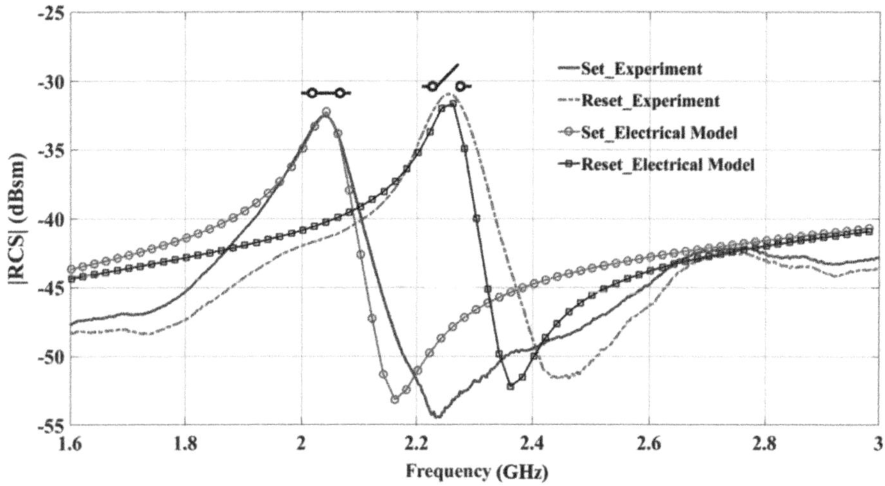

Figure 3.18. *Response of electrical equivalent model of Tag 1 from section 3.3.1, with experimental results. For a color version of this figure, see www.iste.co.uk/jayakrishnan/technology.zip*

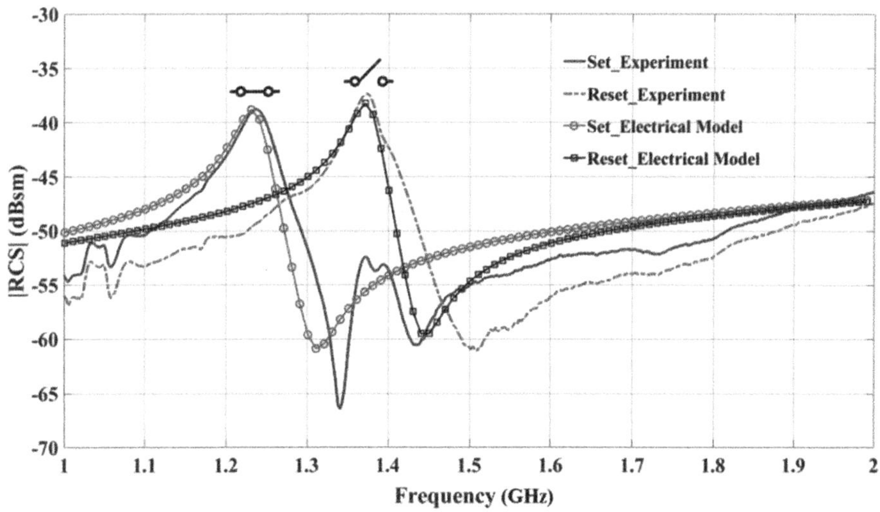

Figure 3.19. *Response of electrical equivalent model of Tag 2 from section 3.3.1, with experimental results. For a color version of this figure, see www.iste.co.uk/jayakrishnan/technology.zip*

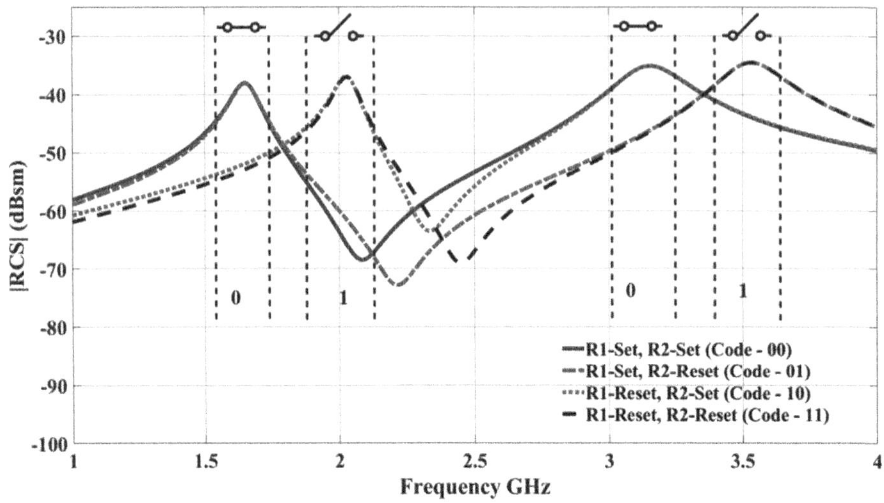

Figure 3.20. *Response of electrical equivalent model of electronically rewritable chipless RFID tag on paper substrate from section 3.3.2. For a color version of this figure, see www.iste.co.uk/jayakrishnan/technology.zip*

The mechanism of operation of the electronically rewritable chipless RFID tags, using "C"-shaped REP with integrated MIM switches as proposed in this chapter, can be explained with the help of the presented electrical model in Figure 3.17. This is described in detail below.

In this explanation, for simplicity we consider only one of the resonators present on the tag, for example resonator 1 (or the antenna mode network of resonator 1 to be more precise). Here, during the set (ON) state of MIM switch, the filament resistance (R_{MIM1}) across the switch is a very small value of the order of 0.5–10 Ω. This is in effect similar to a no resistance path, and thus it short circuits the current path through C_{MIM1}, exhibiting set (ON) state resonance, which depends on the parameters $R_{A1} + R_{MIM1}$, L_{A1} and C_{A1}, where R_{A1}, L_{A1} and C_{A1} are the natural parameters of the metallic resonator, and this resonance is similar in effect to absence of an integrated MIM switch, retaining a short circuit at its place. In the reset (OFF) state of the switch, filament resistance R_{MIM1} is a large value around a few MΩ, similar to an open circuit. This leads to reconstitution of C_{MIM1} to be the major current path in this branch. Hence C_{MIM1} combines in series with C_{A1}, thereby lowering the net capacitance of the resulting RLC combination of this antenna mode network and shifts the resonance to a higher frequency. In all cases, the structural mode network being unchanged has a similar effect in both set (ON) and reset (OFF) states of the integrated MIM switch and the net frequency response of the resonator, which is a combination of the antenna mode and structural mode, is tuned depending on the MIM switch state and the final values of the antenna mode network.

3.6. Discussion of data encoding strategies for electronically rewritable chipless RFID tags based on CBRAM/MIM technology

Section 3.5 presented the electrical model and explained the working mechanism of non-volatile CBRAM/MIM switches, in the proposed rewritable resonators (REPs) for chipless RFID tag applications. From this presentation, we can observe the key difference of a typical MIM switch from an ideal switch, which is the presence of significant OFF or reset (high impedance) state capacitance. Now, in this section, we investigate the possible methods and ideas for utilizing this non-ideal behavior of MIM switches for encoding data in chipless RFID tags.

One could easily understand from analyzing Figure 3.17 and the associated explanations that the reset (OFF) state resonance frequency of the presented REPs is a function of the reset (or high impedance) state capacitance C_{MIM} of integrated MIM switches. Now we present the result of simulation of a rewritable "C"-shaped resonator computed using electrical equivalent circuit in Figure 3.17, for varying C_{MIM} values for set (ON) and reset (OFF) states. We choose the geometry R1 in section 3.3.2 for this calculation. Here, C_{MIM} is varied from 0.5–3 pF and the result is depicted in Figure 3.21, and Figure 3.22 represents a resonance frequency map of variation of resonance frequency as a function of R_{MIM} and C_{MIM}. Readers should note that R_{MIM} in set (ON) state is 1 Ω, and in reset (OFF) state is 1 MΩ for all capacitance states in Figures 3.21 and 3.22.

One could see from Figure 3.21 the dependence of reset (OFF) state resonance of an REP with integrated MIM switch on C_{MIM}, whereas set (ON) state resonance is purely dependent on the geometry of the resonator. Here, a decrease in C_{MIM} from 3 to 0.5 pF induces a difference in displacement of 170–860 MHz among the reset (OFF) state from the set (ON) state resonance frequency of the REP under consideration.

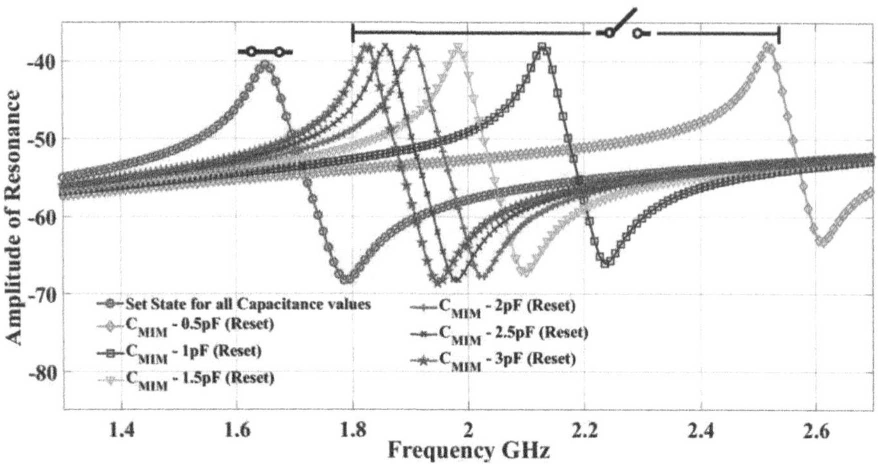

Figure 3.21. *Variation of resonance frequency of an electronically rewritable resonator with integrated MIM switch as a function of* C_{MIM} *and computed using electrical model. For a color version of this figure, see www.iste.co.uk/jayakrishnan/technology.zip*

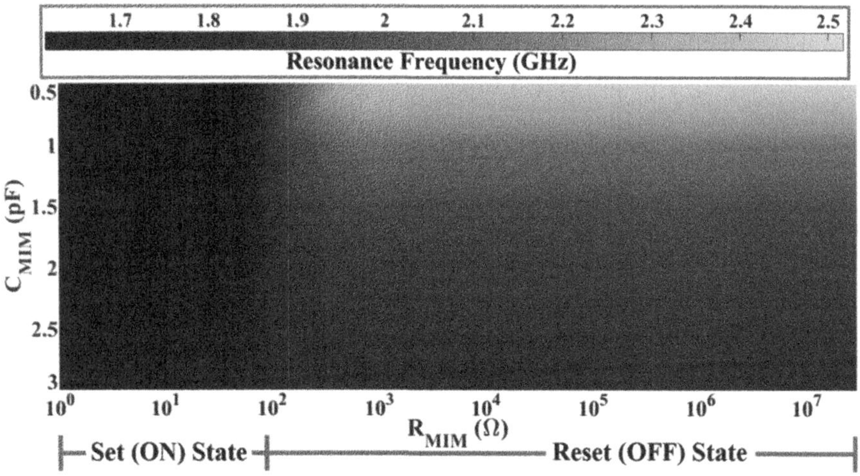

Figure 3.22. *Resonance frequency map of rewritable resonator for variation of* C_{MIM} *from 0.5 to 3 pF and computed using electrical model. For a color version of this figure, see www.iste.co.uk/jayakrishnan/technology.zip*

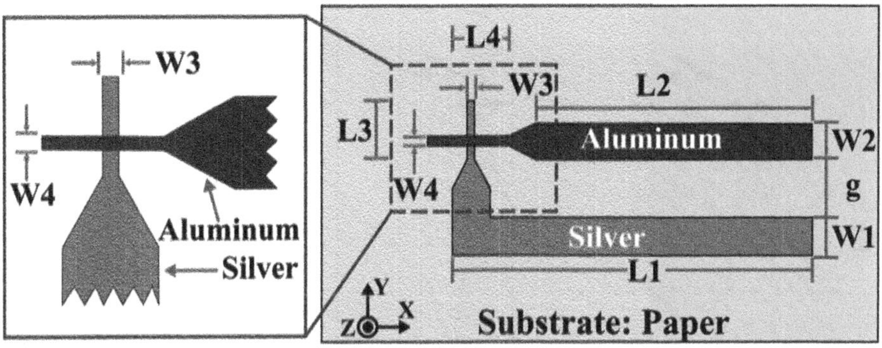

Figure 3.23. *Geometry of electronically rewritable resonator for chipless RFID tag*

To make a judicious utilization of this effect, several options could be derived using this phenomenon for encoding data in a genre of rewritable chipless RFID tags based on the value of achievable C_{MIM} of the integrated MIM switch. Here, C_{MIM} depends on the electrode area of the integrated MIM switch, formed by the intercross section of widths "w3" and "w4"in Figure 3.23 (which is reproduced from Figure 3.4) sandwiching an ion conductor layer, for a given dielectric constant and thickness (which is also the interelectrode gap) of the ion conductor that constitutes a parallel plate capacitor. Resolution of the chosen fabrication process is the key factor that determines this range of C_{MIM} values. Keeping in mind these relations, two simple examples are described below using a tag with "N" C-shaped resonators "R_1" to "R_N" based on the value of C_{MIM} that could be realized.

The first idea is a crisscross arrangement, and it is used when a low-resolution fabrication process is used, like the one presented in this book, with which the achievable trace widths of MIM electrodes are generally greater than hundreds of micrometers. This method of coding is illustrated in Figure 3.24. Here, the reset (OFF) state C_{MIM} is significant (around or greater than 1 pF) due to larger electrode areas and hence the frequency displacement between set (ON) and reset (OFF) states is comparatively small for an REP (e.g. $\Delta F \approx 170$ MHz in the presented experiment). Thus, for this method, the resonances of adjacent resonators for set (ON) and reset (OFF) states are interleaved in a given bandwidth of interest.

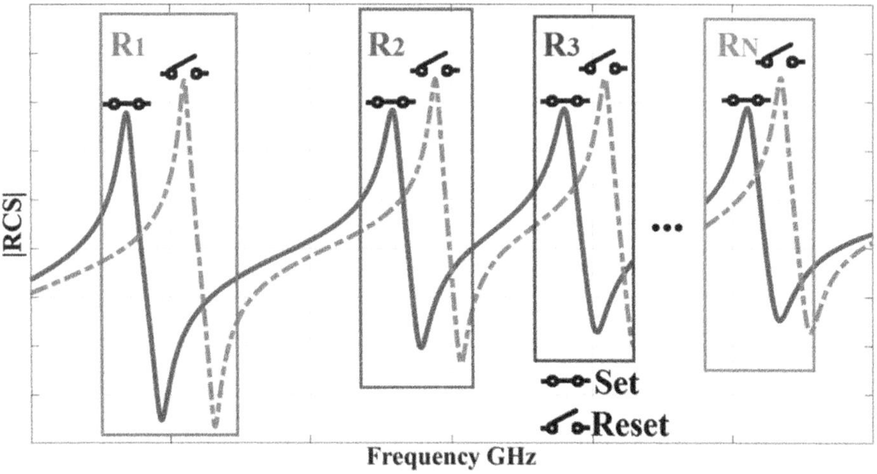

Figure 3.24. *"Crisscross" arrangement of resonance frequencies in an electronically rewritable chipless RFID tag. Encoding proposed for rewritable chipless RFID design, with high C_{MIM} values in the reset (OFF) state of the MIM switch*

Next is a "tuned-out" arrangement that is used when small C_{MIM} values of the order of 0.5 pF and below can be realized using the chosen fabrication process, with a considerable resolution, with which narrow and precise metallic traces (electrodes) having a width of the order of a few tens of micrometers or lower could be achieved. Such REPs are expected to show a considerably large displacement between the set (ON) and reset (OFF) states of the integrated MIM switch (e.g. a displacement $\Delta F \geq 800$ MHz in the presented experiment). Figure 3.25 presents this strategy of coding. Here, the reset (OFF) state resonance frequencies of the tag are tuned away completely from the set (ON) state resonances in the bandwidth of interest.

In both type of coding possibilities presented herewith, an electronically rewritable chipless RFID tag with "N" resonators could represent a minimum of 2^N combinations, combining both set (ON) and reset (OFF) states. Coding capacity could be further increased by adding a greater number of MIM switches per resonator (REP).

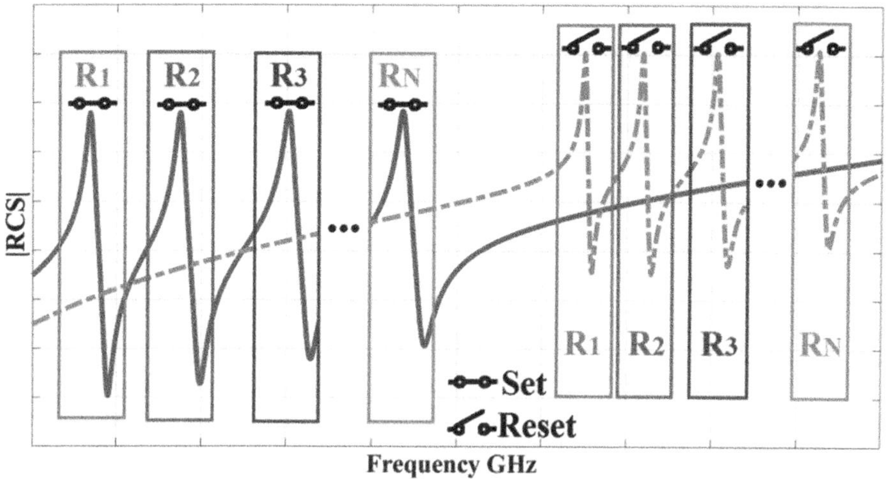

Figure 3.25. *"Tuned out" arrangement of resonance frequencies in an electronically rewritable chipless RFID tag. Encoding proposed for rewritable chipless RFID design, with low* C_{MIM} *values in the reset (OFF) state of the MIM switch*

3.7. Advantages of using integrated CBRAM/MIM switches for chipless RFID applications

In this section, we discuss the advantages of using solid-state electronically rewritable resonators for chipless RFID applications. Before examining the details, it is relevant to discuss the frequency shift coding technique for chipless RFID tags, which is proposed to be used herewith, along with one of the data encoding strategies explained in section 3.6.

Figure 3.26 shows the concept of frequency shift coding (Vena *et al.* 2016; Jayakrishnan *et al.* 2018a) of REPs in a chipless RFID tag. In this technique, generally a narrow band resonance frequency peak (from a given REP) at a given frequency represents a code, and this peak is shifted in frequency in a given bandwidth of interest to different positions to represent other codes. Figure 3.26 shows a chipless RFID tag with a single REP. A "C"-shaped resonator is used as the REP in this case. In this technique, the geometric dimension of REP is tuned in order to achieve a desired frequency of resonance usually referred to as a "spectral signature", which represents a bit or a combination of bits among a group of REPs on a chipless RFID tag, in a given bandwidth.

In general, the coding capacity "N" per tag of a chipless RFID tag utilizing frequency shift coding is defined by equation [3.2] (Jayakrishnan *et al.* 2018a; Vena *et al.* 2016). In equation [3.2], Δf is the allocated bandwidth per REP, df is the bandwidth of resonance of each REP, k is the number of REP on the tag and N is the total number of combinations that are possible. Here, we assume that all REPs have more or less the same bandwidth of resonance.

$$N = \left[\frac{\Delta f}{df}\right]^k \quad [3.2]$$

Classically, modification of the spectral signature, or tuning of a tag, could be done by modifying the dimensions of each REP on the tag physically, as shown in Figure 3.26, using techniques such as micro-machining, laser cutting or inkjet printing. However, such methods are very time consuming and add to the final cost of the tag. In addition, such techniques are mostly one-time modifications, which are usually irreversible.

On the other hand, interestingly, if we introduce a non-volatile CBRAM/MIM switch to this REP as shown in Figure 3.26, the same performance could be achieved by electronically configuring the switch. This kind of electronic rewriteablity is an attractive feature for chipless RFID applications. With this, a tag could be electronically written to configure a tag ID, and then electronically rewritten several times as desired to change the ID, in its life time. Such a feature is not available even with the often referred counterpart of chipless RFID technology the optical barcode system, which is one of the most popular objects or product identifiers used at present. The advent of the Internet of Things (IoT) entails a platform of great demand for such object identifiers, like electronically rewritable chipless RFID tags. In addition, a lot of research has detailed the sensing capabilities of chipless RFID (Nair *et al.* 2013; Amin *et al.* 2014a, 2014b; Feng *et al.* 2015; Borgese *et al.* 2017; Perret 2017). Combining these, we can conceive of a smart label having both sensing and rewritable object identifier capability, which could be used for IoT applications using the proposed techniques in this book.

A proposed application of such a "smart label", as an example to state immediately, would be a low-cost rewritable chipless RFID tag as a multiple target humidity and gas concentration monitor. Such a tag could be deployed as a low-cost smart label in supermarkets and storage places as a food quality monitor for easily perishable packaged food products such as dairy products, meat products and so on. A rewritable chipless RFID tag using the technology presented here could be combined with a humidity sensing capability and various gas sensing capabilities and could be deployed as a label for food packages. Sensing capabilities could be added to the tag through several smart sensing materials (for example, biopolymers for sensing humidity) (Nair *et al.* 2013; Amin *et al.* 2014b; Borgese *et al.* 2017; SAGGIN *et al.* 2019) on a dedicated resonator and monitoring performance parameters such as frequency shift or RCS amplitude, benchmarked against a control experiment. A group of resonators on this tag could be reserved for its ID and others could be dedicated to sensing. Such tags could be deployed for action inside a package, whether visible outside or not. These tags could be monitored either continuously by a centralized reader or by a portable handheld reader upon demand, to display the identity of a packet and any spoilage of food indicated by humidity and gas levels inside the packages. Such "smart labels" could be at the same time identifiers and also sensors of quality; they could even be used for automatic stock updating and for tracking and localization applications (for instance to indicate on which part of the storage shelf they are stored).

Electronically rewritable sensing chipless RFID tags like those proposed here could be fabricated rapidly through a manufacturing process as explained in this book, further optimized for a roll-to-roll industrial process for mass fabrication, as single copy, and then written/rewritten to encode an ID and deployed for use. Advantages of the chosen electronic rewriteablity feature using non-volatile CBRAM/MIM switches, and improvements in convenience using the proposed fabrication technique in this book, are explained in detail below.

In summary, such a label would be of great advantage as a zero power device, with multiple features, for IoT applications where the power supply needs are viewed very critically in terms of stand-alone applications. A passive smart label composed of sensors and identifiers based on passive chipless RFID would be a "deploy and forget" solution. These labels would work on their own, responding to a centralized transponder, and do not bother the user with the need to recharge or change the batteries regularly, being a batteryless solution. We humbly believe that the proposed technology has great potential for future applications as discussed in this section. Further investigation and research for better optimization of the proposed technology is being worked on with utmost seriousness.

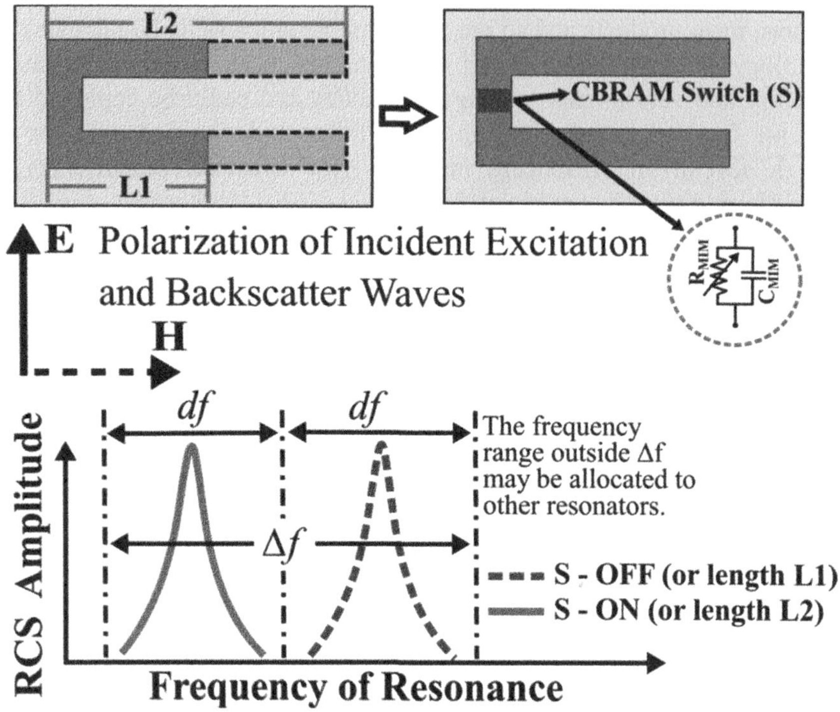

Figure 3.26. *Concept of frequency shift coding used in chipless RFID tags, and advantage of using a CBRAM/MIM switch for electronic rewriteablity*

The expected advantages of using the proposed integration technique of non-volatile CBRAM/MIM switches with chipless RFID tags can be explained through an example. Consider a 20-bit chipless RFID tag utilizing an REP similar to that shown in Figure 3.26 (Vena *et al.* 2012b) for labeling applications using a frequency shift coding technique. Tuning windows of a bandwidth of 200 MHz are allocated to each resonator, and suppose that each resonator has a resonance bandwidth of 100 MHz. This gives a choice of two resonance frequencies for each resonator within the allocated bandwidth. Now if we calculate the possible number of combinations with the tag under consideration using equation [3.2], we can see that we have 1,048,576 possible unique coded tags. Using classic reconfiguring technology like that reported by Vena *et al.* (2013a) and Zheng *et al.* (2008), the following methods are possible. One is to use techniques like photolithography or inkjet printing to fabricate each of the 1,048,576 unique tags. This is a however not a realistic idea. The next practical technique is to use photolithography or a similar technique to realize the whole combination of tags with a single raw dimension and to configure each REP on each tag for a unique code using methods such as micromachining, laser cutting, inkjet printing and so on. In this case, each tag undergoes a unique mechanical process and thus requires unique attention before it can be deployed for application in a real scenario, and in most cases this is an irreversible or write only configuration. In addition, all this precise mechanical process should be failproof and should be done within a "narrow bottleneck" of allowed fabrication tolerance to not degrade the quality of tags due to detuning from desired performances.

On the other hand, if an MIM switch is integrated into each resonator, one has the following advantage. In this case, all 1,048,576 tags could be fabricated as generic copy tags, on any chosen substrates (optimized in design) using the three-step process explained in Chapter 2 for realization of CBRAM/MIM RF switches on flexible substrates, using thermal vapor deposition of metal layers and spin coating of nafion. Each tag fabricated like this would be a blank tag to which any desired code of the base genre could be written or rewritten onto. An electronic write/rewrite stage could also be added in addition to the fabrication process to readily write the desired code to a given set of tags. Otherwise, if the application demands it, a portable electronic programmer similar in size to a credit card reader could be used to electronically write or rewrite each tag at the time of application. The specialty of the proposed write/rewrite technique is that it could be done directly by the customer at the time of application, and tags configured this way would be rewritable at any desired time.

Hopefully, with the confidence from current progress, in the near future these electronically rewritable chipless RFID labels will reach such a state of development that they can be printed and programmed in the field by the customer. Printing of the tags could be done using an ordinary inkjet printer using conductive inks, and ion conductor resin solutions, thus making a complete "electronically rewritable RF barcode". The proposed rewritable RF labels can also be used for low-cost secure counter applications reserving a specific number of bits for identification and the rest for counting, for applications like in disposable public transport tickets. An illustration of the proposed concept for an electronically rewritable chipless RFID label is shown in Figure 3.27, along with an example of a portable tag programmer.

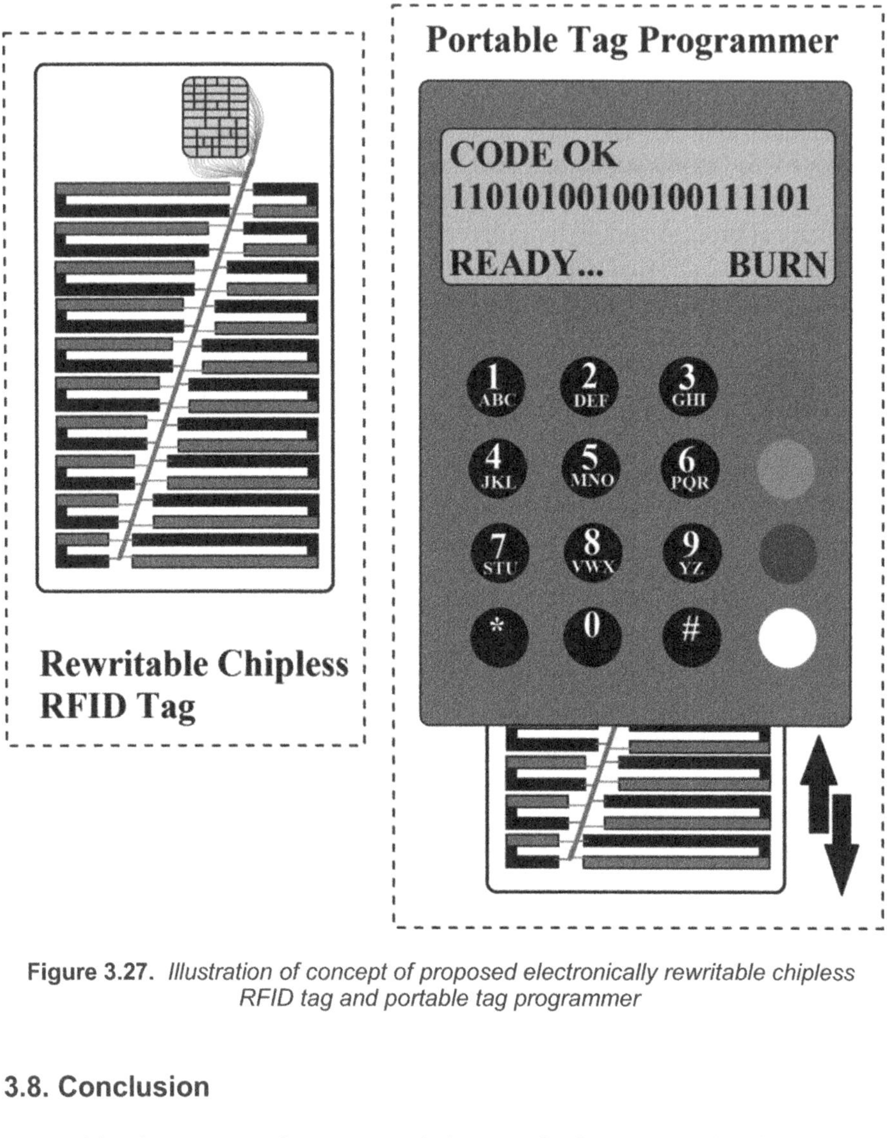

Figure 3.27. *Illustration of concept of proposed electronically rewritable chipless RFID tag and portable tag programmer*

3.8. Conclusion

In this chapter, we demonstrated the proof of concept of utilization of solid-state and non-volatile CBRAM/MIM switches for electronically rewritable chipless RFID applications. The working mechanism of rewritable resonators (REPs) for chipless RFID tags was explained with the help of an electrical equivalent model. We also presented possible coding strategies and highlighted the advantages of using the proposed techniques for commercial labeling applications. We can say with the confidence obtained from these experimental results that the commercial production of fully printed electronically rewritable RF-barcode stickers is not very far in the near future. This would be a perfect solution for various wireless identifier requirements for the IoT and also for other commercial and domestic applications.

The use of CBRAM/MIM RF switches could be extended to most switching requirements of low-cost and passive RF and microwave devices. In this respect, the application of CBRAM/MIM switches to electronically pattern steerable antennas, frequency reconfigurable filters and their analysis are discussed in detail in Chapter 4.

4

Fully Passive Solid-State Electronically Reconfigurable Filter and Antenna Models

4.1. Introduction

In this chapter, we present the applications of the proposed conductive bridging random access memory/metal insulator metal (CBRAM/MIM) radio frequency (RF) switches for electronically reconfigurable filters and antennas. We try to demonstrate the potential applications of CBRAM/MIM switches for enhancing the performance of classic RF and microwave devices. In order to highlight the advantages of integrating CBRAM/MIM switches into passive RF and microwave devices, we examine the generalized design topology of a band-stop filter and a parasitic element (PE)-loaded dipole antenna. We strongly believe that the reader will be convinced, by the end of this chapter, of the conclusive advantages of non-volatile reconfigurable RF devices based on the proposed switch technology, in contrast to their vintage predecessors, for low-power stand-alone applications.

4.2. CBRAM-MIM switches for electronically reconfigurable filter applications

Multiband reconfigurable filters would enormously benefit the microwave and wireless communication of the present era (Hong 2009; Al-Yasir *et al.* 2019). Electronically reconfigurable multiband RF filters would support reductions in the physical volume and signal processing budget of a system, contributing to increased versatility. Ample research has been carried out in this domain and is still advancing at a good rate, and a handful of remarkable prototypes have been recently presented (Hong 2009; Chaudhary *et al.* 2012, 2013; Psychogiou *et al.* 2016; Feng *et al.* 2017; Chen *et al.* 2017; Al-Yasir *et al.* 2019).

However, in simple terms, most of these prototypes utilize an active switching technology such as PIN diodes, other solid-state semiconductor switches, or micro-electromechanical system (MEMS) switches to achieve reconfigurability characteristics. These switch technologies in general require a power source to maintain ON/OFF impedance states, and are hence active and volatile devices. Moreover, such switches are added to the target device, in most cases, by intrusive techniques like soldering, which tend to degrade RF performance of the RF circuit to some extent.

In this chapter, we demonstrate the design and experimental realization of an electronically reconfigurable band-stop filter using non-volatile CBRAM/MIM switches directly integrated to a planar filter topology, notably without any techniques like soldering. Such an electronically reconfigurable filter design using polymer ion conductor-based non-volatile CBRAM/MIM switches was initially reported in Jayakrishnan (2019b). This realization was done through simple fabrication processes as explained in Chapter 2 for CBRAM RF switches, without utilizing any "clean room" techniques. We also present the extension of the proposed technology to more efficient filter topologies using full-wave electromagnetic simulation studies.

4.2.1. *Electronically reconfigurable band-stop filter*

A simple design for an electronically reconfigurable band-stop filter to demonstrate the application of CBRAM/MIM switches for such applications is presented herewith. Here, two filter topologies named design N1 and N2, with similar working mechanisms, are realized. The band-stop filters are based on tuned stubs loaded on to a microstrip line, with integrated CBRAM/MIM switches for electronic reconfigurability. Here, integrated CBRAM/MIM switches tune the electrical length of the stub forcing it to resonate at two different frequencies, respectively, for set (low impedance) and reset (high impedance) states.

Figure 4.1 shows the proposed topology of the filters. This design consists of a high impedance stub attached to a classic 50 Ω microstrip transmission line. A CBRAM/MIM switch is integrated into the stub by tapering its width as shown in Figure 4.1. The CBRAM switch is formed by sandwiching a 600 nm nafion layer (ion conductor), among the metal traces marked by width "w3" (copper) and "w4" (aluminum), which form the active and inert electrodes, respectively.

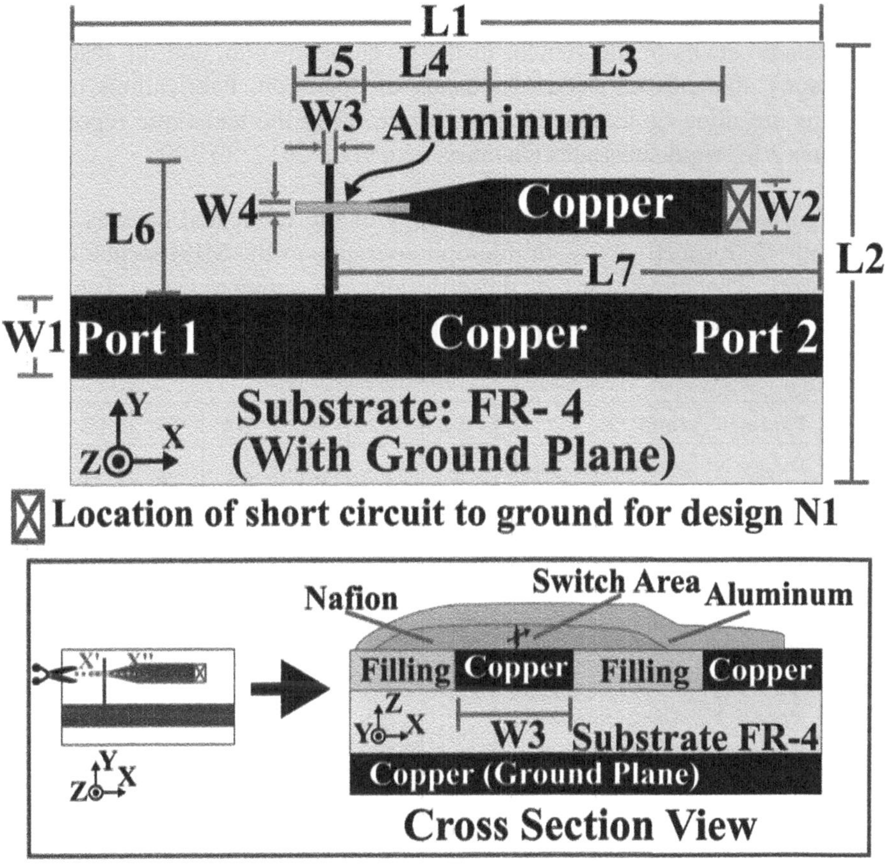

Figure 4.1. *Topology of proposed electronically reconfigurable band-stop filter. Inset shows cross-section view of layer structure of MIM switch along the plane x' – x". Layer thickness: FR-4: 1.6 mm, copper: 35 μm, aluminum: 500 nm–1.5 μm, nafion: 600 nm. Refer to Table 4.1 for dimensions of design N1 and N2. For a color version of this figure, see www.iste.co.uk/jayakrishnan/technology.zip*

Design N1 is a shorted stub-based band-stop filter, in which the stub is short circuited by connecting it to the ground plane of one of the two SMA connectors at the input/output ports as shown in Figure 4.2; N2 is an open stub-based band-stop filter, as shown in Figure 4.3. The stop band of design N1 could be switched OFF or ON depending, respectively, on the set (low impedance) or reset (high impedance) state of the integrated CBRAM switch. The short-circuited stub is resonant in the bandwidth of interest only in the reset (high impedance) state of the switch. However, for N2, the open circuited stub is resonant for both set (low impedance) and reset (high impedance) states of the switch, in the given bandwidth, and the resonance frequency switches between two bands of operation. Fabrications of both designs are done on classic FR-4 substrate using the technique reported in Chapter 2 for rigid substrates (Jayakrishnan *et al.* 2017, 2019b).

Figures 4.2 and 4.3 show photographs of the fabricated devices named N1 and N2, respectively, with microphotographs of the MIM switch area in the insets. The dimensions of these designs are summarized in Table 4.1. Better insight into the mechanism of operation is explained in detail with the help of an electrical model in section 4.2.1.2.

Parameters (mm) →	**W1**	**W2**	**W3**	**W4**	**L1**	**L2**
Design N1 (shorted stub-based band-stop filter)	3	2	0.1	0.3	28	18
Design N2 (open stub-based band-stop filter)	3	2	0.1	0.3	40	18
Parameters (mm) →	**L3**	**L4**	**L5**	**L6**	**L7**	–
Design N1 (shorted stub-based band-stop filter)	3.85	5	2.6	5	14	–
Design N2 (open stub-based band-stop filter)	14.7	5	2.6	5	30	–

Table 4.1. *Dimensions of electronically reconfigurable band-stop filters for geometry shown in Figure 4.1 for design N1 and N2*

Proposed design is initially optimized using full-wave electromagnetic simulation in CST Microwave Studio using a lumped resistor to model set (low impedance) and reset (high impedance) states of the switch, as explained in Chapter 2, for the CPW RF switch. Photo-lithographically engraved copper clad of FR-4 PCB is used as the active electrode, and a layer of thermal vapor deposited aluminum forms the inert electrode. Nafion resin solution is spin coated to achieve a thickness of 600 nm, to form the ion conductor layer, sandwiched among the electrodes marked by width "w3" and "w4" in Figure 4.1.

Integrated CBRAM switches in the filters are operated using DC pulses, and are applied directly for N1 through SMA connectors and for N2 using ordinary crocodile clips connected to the respective metallic parts. A triangular pulse of peak voltage +16 V, with a compliance current limit of 0.5 mA, is used for the set process (low impedance state) and a negative square pulse of –20 V with a compliance current limit of 150 mA is used for the reset process (high impedance state) of CBRAM switch, as explained in Chapter 2 (section 2.2.1).

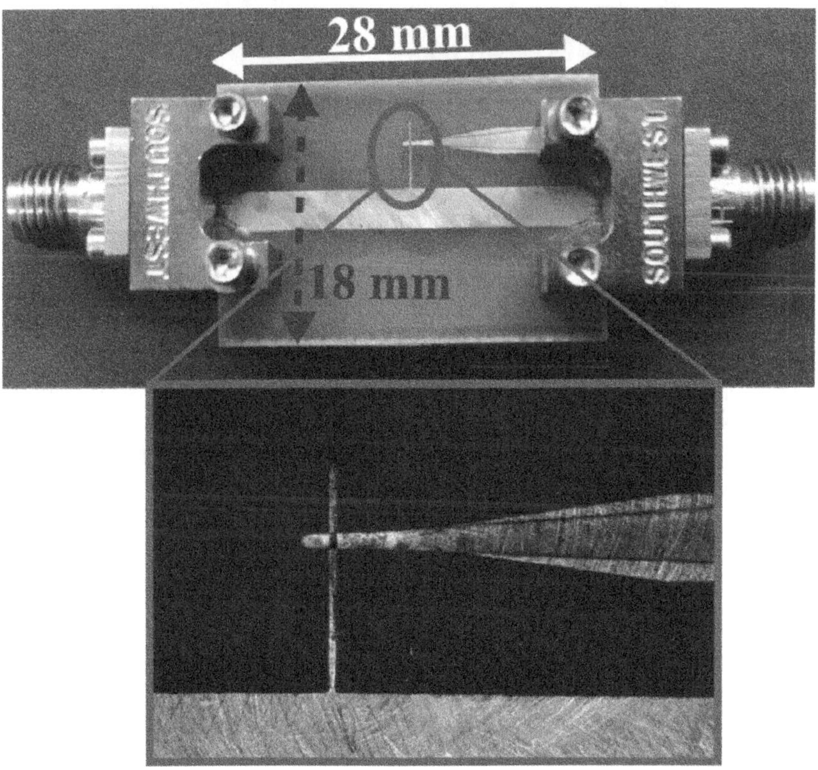

Figure 4.2. *Photograph of fabricated electronically reconfigurable shorted stub-based band-stop filter (N1). Inset shows microphotograph of MIM switch area*

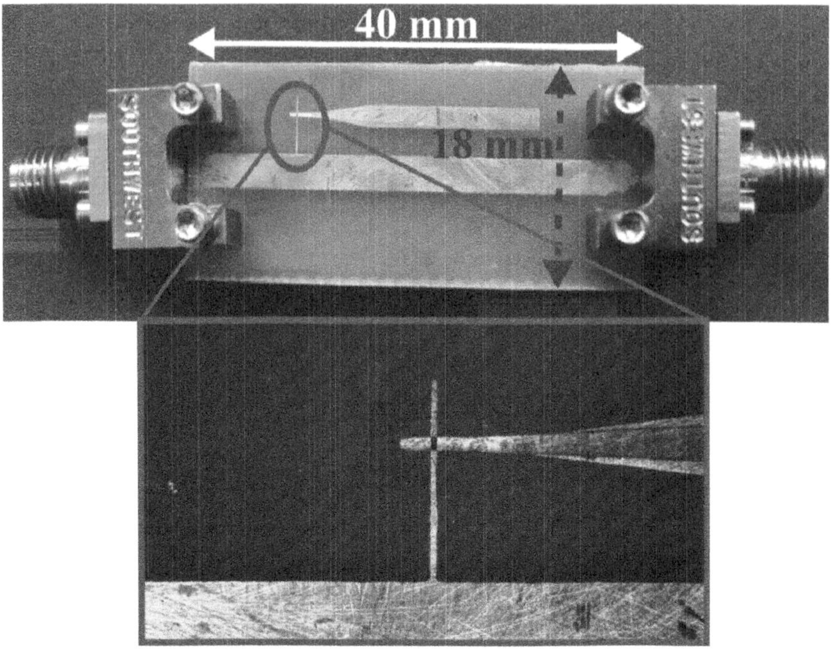

Figure 4.3. *Photograph of fabricated electronically reconfigurable open stub-based band-stop filter (N2). Inset shows microphotograph of MIM switch area*

4.2.1.1. *Experimental results*

RF response of these realized filters were recorded using Agilent ENA E5061B Network Analyzer using a classic Short-Open-Load-Thru (SOLT) calibration scheme. Observed experimental RF response of the filters along with results of full-wave electromagnetic simulation are as follows.

Figures 4.4 and 4.5 shows the S21 and S11, respectively, for N1 (shorted stub-based band-stop filter). Figures 4.6 and 4.7 show the S21 and S11, respectively, for N2 (open stub-based band-stop filter). We can observe from these plots that the results obtained in experiment are in good agreement with the results of full-wave electromagnetic simulations.

For design N1, the stop band is centered at 1.5 GHz and shows ON/OFF switching action depending on set/reset sates of the MIM switch. Passband insertion loss less than –2 dB and stop band isolation greater than –11 dB, with –10 dB bandwidth of 80 MHz, is observed for this design. Similarly, for N2 the set (low impedance) state resonance is centered at 1.42 GHz and is shifted by 590 MHz to 2.01 GHz in the reset (high impedance) state of the MIM switch. N2 shows an insertion loss less than –2 dB in passband, and isolation greater than –14 dB in stop band. Design N2, in both stop bands, exhibits a –10 dB band width greater than 140 MHz.

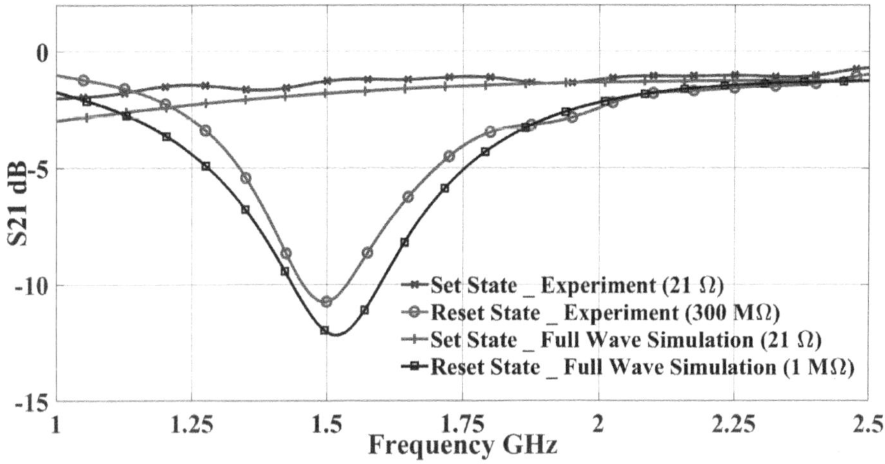

Figure 4.4. *Experimentally obtained S21 response of electronically reconfigurable shorted stub band-stop filter (N1). For a color version of this figure, see www.iste.co.uk/jayakrishnan/technology.zip*

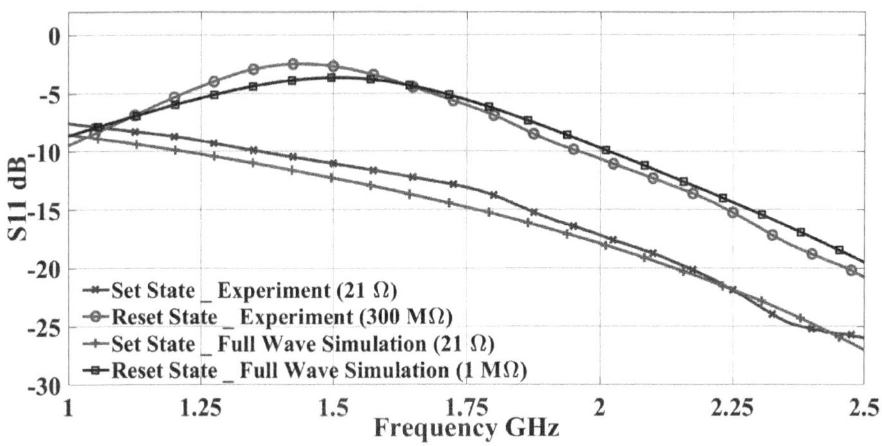

Figure 4.5. *Experimentally obtained S11 response of electronically reconfigurable shorted stub band-stop filter (N1). For a color version of this figure, see www.iste.co.uk/jayakrishnan/technology.zip*

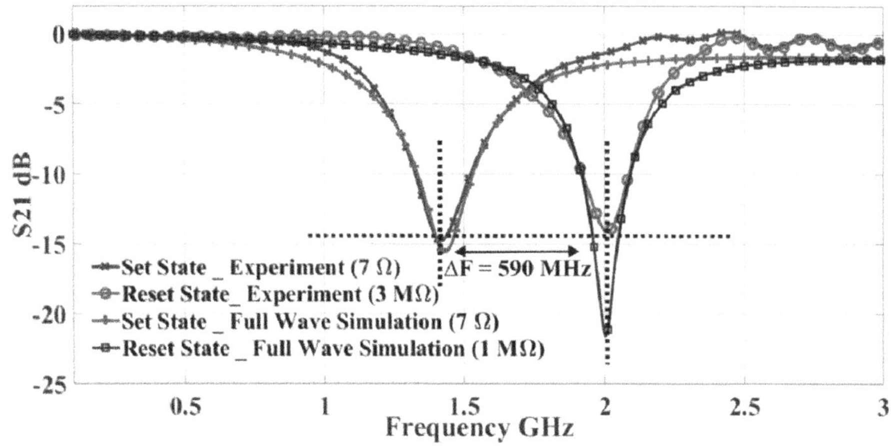

Figure 4.6. *Experimentally obtained S21 response of electronically reconfigurable open stub band-stop filter (N2). For a color version of this figure, see www.iste.co.uk/jayakrishnan/technology.zip*

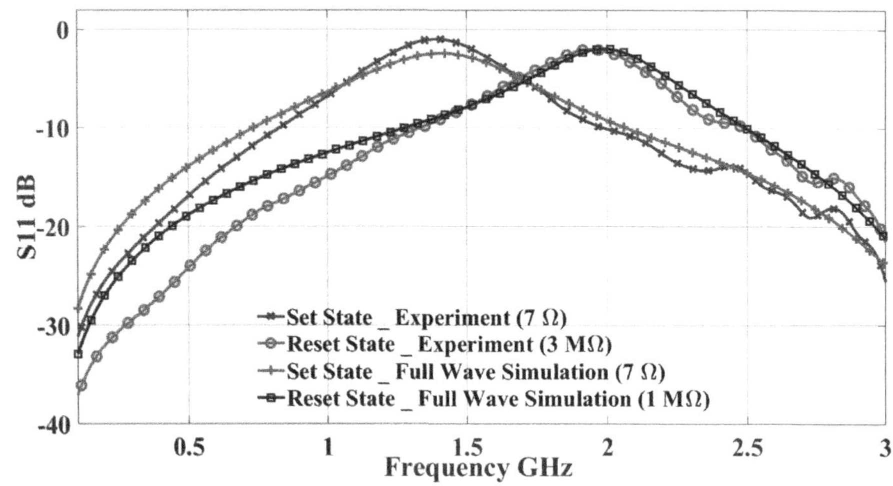

Figure 4.7. *Experimentally obtained S11 response of electronically reconfigurable open stub band-stop filter (N2). For a color version of this figure, see www.iste.co.uk/jayakrishnan/technology.zip*

Set (low impedance) and reset (high impedance) state resistances of MIM switch were experimentally observed to be in the range of 1–30 Ω and 1–300 MΩ, respectively, for most of the cycles for both devices N1 and N2.

The RF results presented here are inclusive of effects of associated transmission lines, SMA connectors and feed lines.

4.2.1.2. *Equivalent electrical model and mechanism of operation*

In this section, we present a simple electrical model to explain the working mechanism of the realized electronically reconfigurable band-stop filters. The presented electrical model is a basic approximation focusing on the effect of the proposed integrated MIM switch in the stub-loaded filters for the principle harmonic frequency of operation. Yet this is believed to be a congenial model in giving the reader a good insight into the mechanism of operation of the presented devices.

Figures 4.8 and 4.13 show, respectively, the electrical equivalent models of design N1 and N2 (treated as separate sections later in this chapter). A CBRAM/MIM switch like the one integrated in these filters could be modeled as an RC parallel network as explained in Chapter 2 (Pi *et al.* 2015; Jayakrishnan *et al.* 2017, 2018a). Next, a resonant stub loaded on to a microstrip line could be modeled as an RLC serial network (Amin *et al.* 2018).

Parameters (design N1)	**Values Design N1 (shorted stub band-stop filter)**	**Parameters (design N2)**	**Values Design N2 (open stub band-stop filter)**
R_{MIM}	21 Ω/1 MΩ*	R_{MIM}	1 Ω/1 MΩ*
C_{MIM}	1.4 pF	C_{MIM}	1.48 pF
R_{SS}	9 Ω	R_{OS}	5 Ω
L_{SS}	8 nH	L_{OS}	8.35 nH
*For set–low impedance and reset–high impedance states of MIM switch.		C_{OS}	1.5 pF

Table 4.2. *Parameter values of electrical equivalent model of electronically reconfigurable band-stop filters in Figures 4.8 and 4.13*

In the presented cases of reconfigurable filters, we can combine these concepts as shown in Figures 4.8 and 4.13 to represent the proposed band-stop filters. Detailed explanations of the model and working mechanism of the filters are explained in the following sections separately for design N1 (section 4.2.1.2.1) and N2 (section 4.2.1.2.2). Component values of electrical equivalent model parameters for both design N1 and N2 are summarized in Table 4.2.

4.2.1.2.1. Design N1 (electronically reconfigurable shorted stub band-stop filter)

The presented reconfigurable shorted stub-based band-stop filter could be approximated as an RL series network composed by R_{SS} and L_{SS} for the shorted stub in series with parallel RC approximation (composed of R_{MIM} and C_{MIM}) for the CBRAM/MIM switch, as shown in Figure 4.8.

In the set (low impedance) state of the switch (or in absence of a switch i.e. when the two arms of the cross marked by w3 and w4 in Figure 4.1 are short-circuited to each other, without the sandwiched ion conductor), the load stub is not resonant in the targeted bandwidth. In reset (high impedance) state, it resonates with a surface current pattern much less than a quarter-wavelength distribution.

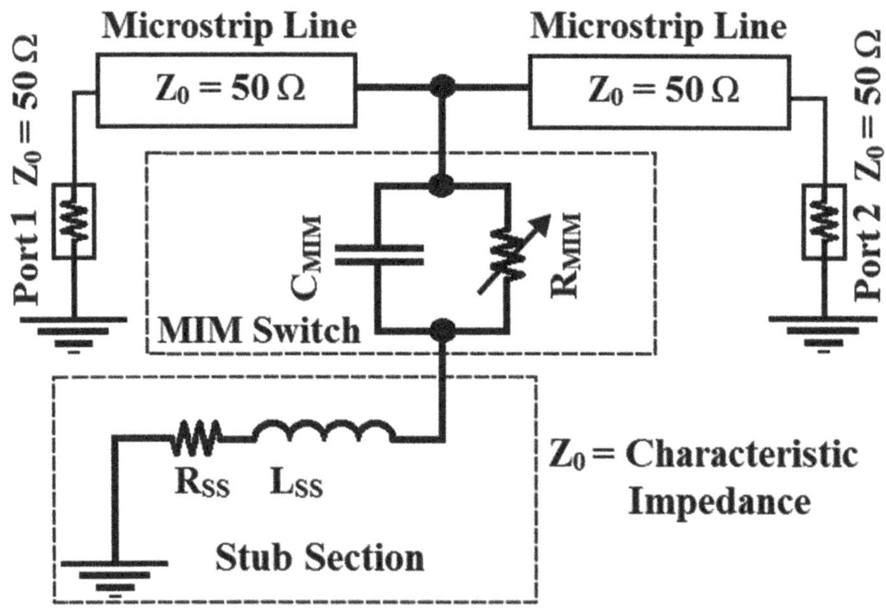

Figure 4.8. *Electrical equivalent model of electronically reconfigurable shorted stub band-stop filter (N1). Refer to Table 4.2 for component parameter values*

The operating mechanism of filter N1 could be explained as follows with respect to the electrical model in Figure 4.8. In the set state of the switch, R_{MIM} is a very low value ($<25\ \Omega$), similar to a short circuit, and hence most of the current in this arm of network flows through this path. This is strong enough to ignore the current path through C_{MIM} and thus the stub constitutes a non-resonant path. Thus not much power flowing through the microstrip line is bypassed to ground through this path due to the high impedance offered by inductance L_{SS} at this frequency. This establishes the all pass filter characteristics of N1.

Inversely, in the reset state of the switch, R_{MIM} is a large value (1 MΩ) similar to an open circuit. This tends to ignore R_{MIM} from the current path, and major current flow would be only through C_{MIM} (or a very small and negligible amount through R_{MIM}) and the stub. This results in a resonant RLC circuit as a combination of C_{MIM}, R_{SS} and L_{SS}. This constitutes a low impedance path through the stub, and most of the power originating from port 1 (or port 2) and flowing through the microstrip line is bypassed to ground, around the resonance frequency exhibiting band-stop characteristics. Surface current plots of design N1 obtained in full-wave electromagnetic simulation are shown in Figures 4.9 and 4.10, respectively, for set (low impedance) and reset (high impedance) states of the CBRAM/MIM switch. Figure 4.9 shows the surface current pattern at 1.5 GHz, which constitutes a pass band in the set state of MIM switch, and Figure 4.10 shows the surface current pattern for the stop band, for the reset state of MIM switch at 1.5 GHz.

The electrical model in Figure 4.8 is optimized using circuit modeling in the schematic mode of CST Microwave Studio. The starting values of R_{MIM} are taken from experimentally observed filament resistances of the integrated MIM cell, and values of C_{MIM} are chosen initially based on calculations from the geometry of the integrated MIM switch. All component parameter values are then later enhanced for the best performance through fitting to experimentally observed electromagnetic characteristics of the filter.

Figures 4.11 and 4.12 show subsequently the S21 and S11 response of the electrical model for design N1 in comparison to experimental results for reference. Here, we can observe the agreement between the two plots that validates the accuracy of the proposed electrical equivalent models.

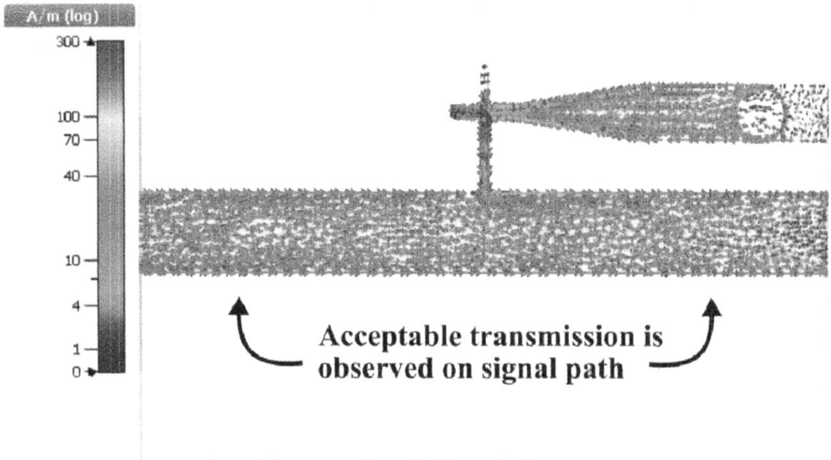

Figure 4.9. *Surface current distribution of electronically reconfigurable shorted stub band-stop filter (N1) at 1.5 GHz, in set (low impedance) state, which constitutes a pass band. For a color version of this figure, see www.iste.co.uk/jayakrishnan/technology.zip*

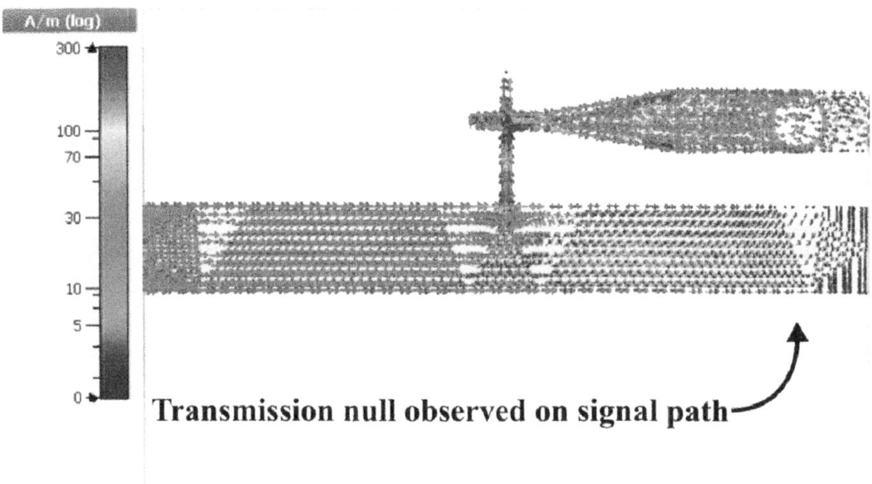

Figure 4.10. *Surface current distribution of electronically reconfigurable shorted stub band-stop filter (N1) at 1.5 GHz, in reset (high impedance) state, which constitutes a stop band. For a color version of this figure, see www.iste.co.uk/jayakrishnan/technology.zip*

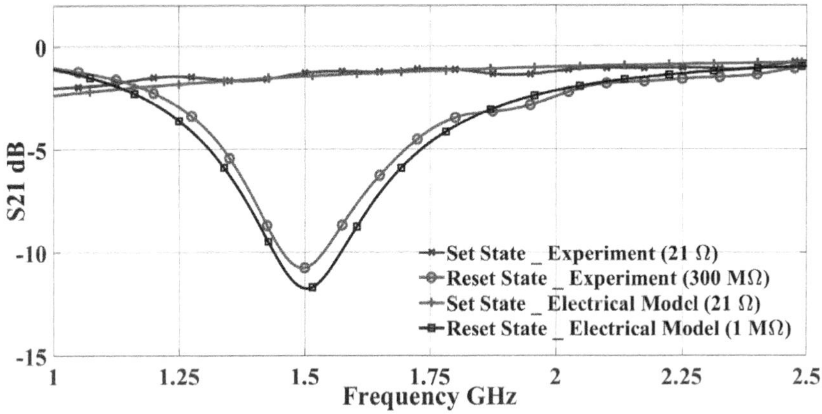

Figure 4.11. *S21 response of electrical equivalent model of electronically reconfigurable shorted stub band-stop filter (N1). For a color version of this figure, see www.iste.co.uk/jayakrishnan/technology.zip*

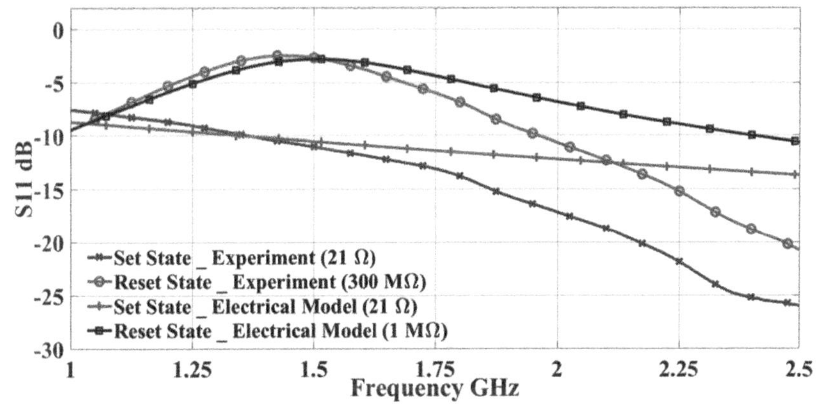

Figure 4.12. *S11 response of electrical equivalent model of electronically reconfigurable shorted stub band-stop filter (N1). For a color version of this figure, see www.iste.co.uk/jayakrishnan/technology.zip*

4.2.1.2.2. Design N2 (electronically reconfigurable open stub band-stop filter)

In this case, the equivalent circuit of design N2 (electronically reconfigurable open stub band-stop filter) could be modeled as a serial combination of a parallel RC equivalent for the CBRAM/MIM switch and an RLC serial network for the open circuit stub, as shown in Figure 4.13.

Surface current patterns obtained in full-wave electromagnetic simulation for this design at resonance show a distribution similar to quarter wavelengths for both states of the MIM switch.

Similar to design N1 presented above, the electrical model in Figure 4.13 is optimized using circuit modeling in CST Microwave Studio. The starting values of R_{MIM} are taken from experimentally observed filament resistances of the integrated MIM cell, and the values of C_{MIM} are chosen initially based on calculations from the geometry of the integrated MIM switch. All component parameter values were then enhanced later for the best performance, through fitting with experimentally observed electromagnetic characteristics of the filter.

Figures 4.14 and 4.15, respectively, show the surface current pattern on the stub, at set (low impedance) state resonance frequency (1.42 GHz), which constitutes a stop band, and at the same frequency, non-resonant in reset (high impedance) state, which constitute a pass band. Surface current plots at 2.01 GHz, which constitute a stop band in reset (high impedance) state and pass band in set (low impedance) state, are also similar and are omitted to maintain brevity.

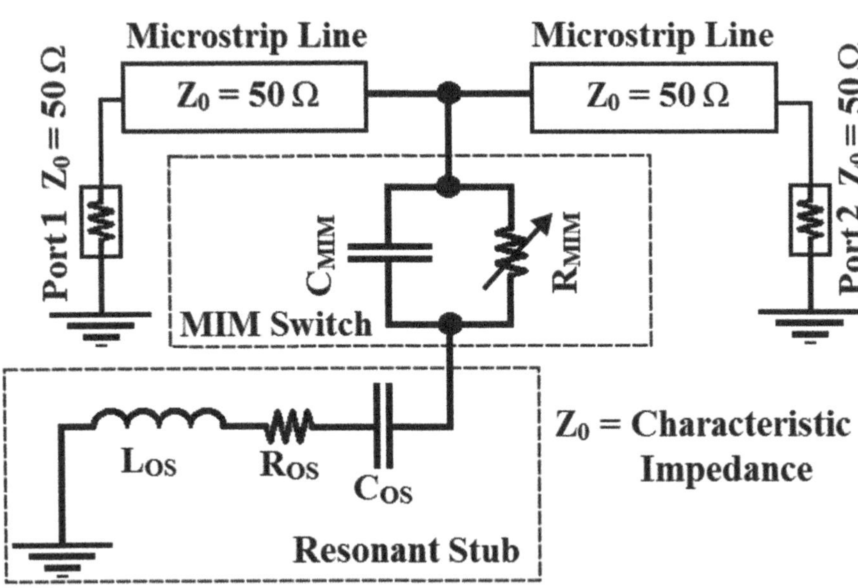

Figure 4.13. *Electrical equivalent model of electronically reconfigurable open stub band-stop filter (N2). Refer to Table 4.2 for parameter values*

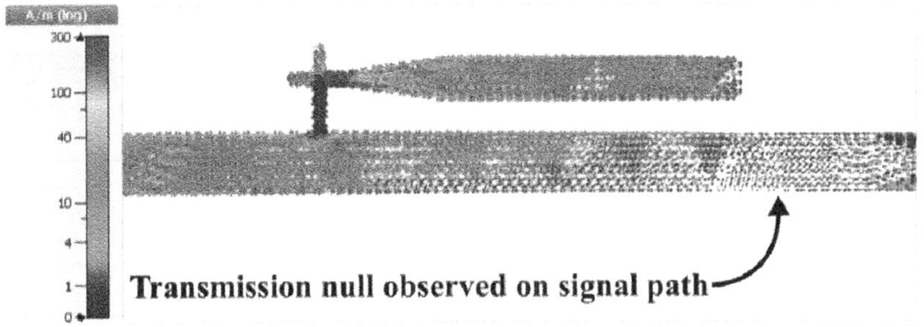

Figure 4.14. *Surface current distribution of electronically reconfigurable open stub band-stop filter (N2) at 1.42 GHz, in set (low impedance) state, which constitutes a stop band. For a color version of this figure, see www.iste.co.uk/jayakrishnan/technology.zip*

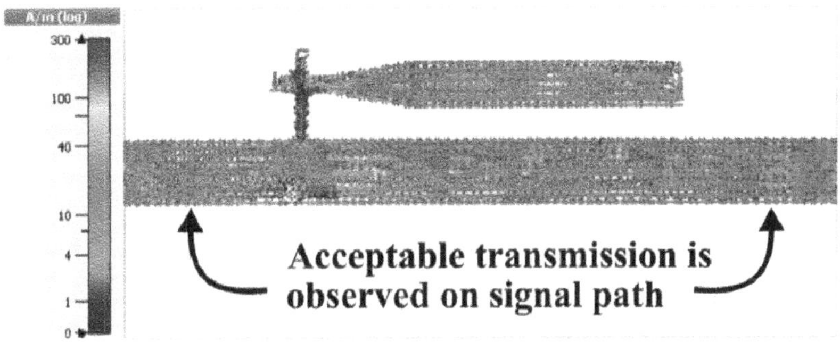

Figure 4.15. *Surface current distribution of electronically reconfigurable open stub band-stop filter (N2) at 1.42 GHz, in reset (high impedance) state, which constitutes a pass band. For a color version of this figure, see www.iste.co.uk/jayakrishnan/technology.zip*

The operating mechanism of N2 is also similar to the above theory explained for N1, for different switch states of the integrated MIM cell. On the other hand, in this case, the filtering stub is long enough to be resonant in both switch states, and is represented by an RLC serial network as shown in Figure 4.13. Thus, a band-stop behavior in both set (low impedance) and reset (high impedance) states of the MIM switch is exhibited by this design, at different frequencies, depending on the directed current path through R_{MIM} or C_{MIM} and resulting net value of capacitance due to the combination of C_{MIM} and C_{OS} of the RLC equivalent of the resonating stub. Here, in the set state of MIM switch, R_{MIM} is a small value (<10 Ω) that effectively short-circuits C_{MIM}, and resonant frequency depends only on R_{OS}, L_{OS} and C_{OS}.

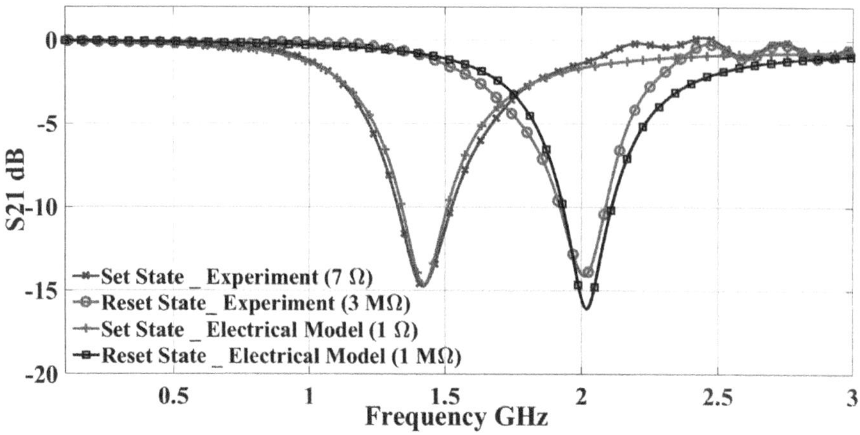

Figure 4.16. *S21 response of electrical equivalent model of electronically reconfigurable open stub band-stop filter (N2). For a color version of this figure, see www.iste.co.uk/jayakrishnan/technology.zip*

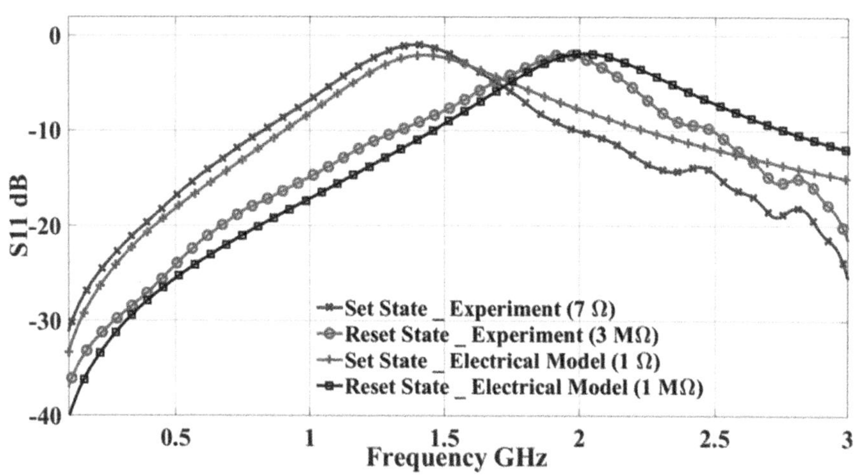

Figure 4.17. *S11 response of electrical equivalent model of electronically reconfigurable open stub band-stop filter (N2). For a color version of this figure, see www.iste.co.uk/jayakrishnan/technology.zip*

Inversely, in the reset state of the MIM switch, R_{MIM} is a large value (1 MΩ) which should be ignored in the current path and hence C_{MIM} is connected to C_{OS} in series, resulting in a lower net capacitance (0.7 pF), and the resonant frequency is shifted to a higher value. Figures 4.16 and 4.17

show the S21 and S11 responses of the electrical model for design N2 in comparison to experimental results for reference. Here, we can observe the agreement between the two plots that validates the accuracy of the proposed electrical equivalent models.

4.2.1.3. *Effect of MIM switch capacitance on resonance frequency of presented electronically reconfigurable filters*

In both filter designs explained above, the interelectrode capacitance C_{MIM} of the integrated CBRAM/MIM cell, in reset (high impedance) state, is seen to exhibit an acute control over the resonance frequencies constituting filtering characteristics. Capacitance C_{MIM} (stated by classic equations for parallel plate capacitance) could be tuned by controlling the surface area of electrodes sandwiching the ion conductor (with a predefined thickness), which forms the integrated MIM cell. These electrodes are marked by width "w3" and "w4", as shown in Figure 4.1. This particular observation in effect could be used to control the displacement between frequencies of resonance in set (low impedance) and reset (high impedance) states for filter design N2. This phenomenon is described in Figure 4.18, which shows the resonance frequency variation of filter N2 as a function of C_{MIM} through a simulation study using the electrical equivalent model.

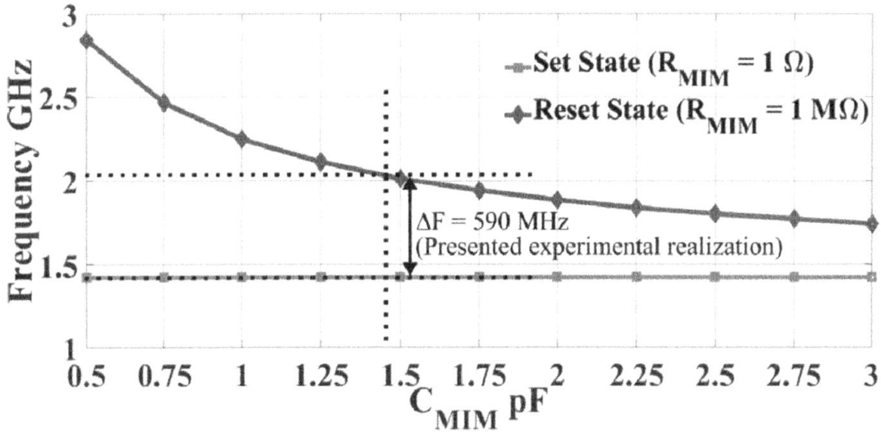

Figure 4.18. *Resonance frequency as a function of C_{MIM} for electronically reconfigurable open stub band-stop filter (N2)*

In this plot, we can also observe a displacement of 590 MHz between the set (low impedance) and reset (high impedance) state resonance frequencies for a C_{MIM} value of 1.48 pF as obtained with the experimental realization of design N2 (see Figure 4.6 and Table 4.2) as presented in this chapter.

In summary, a desired higher or lower separation between set (low impedance) and reset (high impedance) state resonance frequencies, as a result of corresponding reset state capacitance values of CBRAM switch, could be achieved by controlling electrode widths appropriately in design. This is only limited by the least achievable feature size according to the available fabrication technique, which is typically around ~50 µm for the common lithography or thermal vapor deposition process used for fabrication in this study, or a few hundreds of nanometers to few micrometers for "clean room"-based microfabrication. Similarly, for design N1, the resonance frequency for a given length of stub could be fine-tuned to a desired value utilizing this effect.

One should also note that the length "L3" of the stub in Figure 4.1 also could be used to vary the gross resonance frequency, for N2 and N1. The control on C_{MIM} is a great advantage in optimizing the desired filtering characteristics, in such a general design, by not extensively altering the available physical space.

In addition, even though not yet investigated experimentally, the ion conductor thickness of CBRAM switches could also be used for controlling the reset (high impedance) state C_{MIM}, provided it ensures an acceptable set and reset state impedance levels. In this study, this value is kept fixed to 600 nm in order to ascertain appropriate low impedance in set states.

A simple mathematical description of this design strategy is presented as follows. Mathematical relations for determining the frequency of resonance (F_r) of open stub band-stop filter N2 presented in this experiment, which is similar to a quarter wavelength resonator, is given by equations [4.1]–[4.3]. Here, L_{eff} is the effective length of the resonant quarter wavelength stub. Now, L_{eff} could be represented as a function of a tune length (L_t) of the stub and a tune width (w_t) of the MIM switch electrode area for both set (low impedance) and reset (high impedance) states of the integrated MIM switch. "*L3*" and "*w4*" (from Figure 4.1) are the chosen tune length and tune width, respectively, in this design.

$$F_r = \frac{c}{4\, L_{eff}\sqrt{\varepsilon_{eff}}} \qquad [4.1]$$

$$L_{eff} = -0.8w_t + L_t + 15 \quad \text{(For set – low impedance state)} \qquad [4.2]$$

$$L_{eff} = -26w_t^2 + 32.7w_t + 0.5L_t + 6 \quad \text{(For reset – high impedance state)} \qquad [4.3]$$

In [4.1]–[4.3], c represents the speed of light in a vacuum measured in m/s and ε_{eff} represents the effective dielectric permittivity of the chosen substrate, and in the presented case it is calculated to be 3.16 (dimensionless).

Nonlinear multiple variable regression analysis using the method of least squares is used to model these equations with the help of Matlab. Optimization of this model is done using full-wave electromagnetic simulation data and is valid for "L_t" in the range of 10–18 mm, and "w_t" in the range of 0.1–0.7 mm. Figures 4.19 and 4.20 show the resonance frequency values obtained using [4.1]–[4.3], for the above specified range, for set (low impedance) and reset (high impedance) states of the integrated CBRAM/MIM switch.

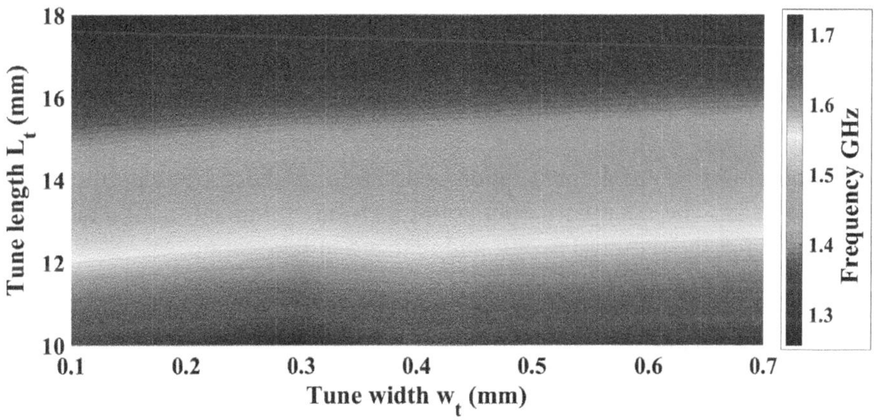

Figure 4.19. *Resonance frequency calculated using [4.1]–[4.3] for set (low impedance) state of CBRAM/MIM switch for electronically reconfigurable band-stop filter design N2. For a color version of this figure, see www.iste.co.uk/jayakrishnan/technology.zip*

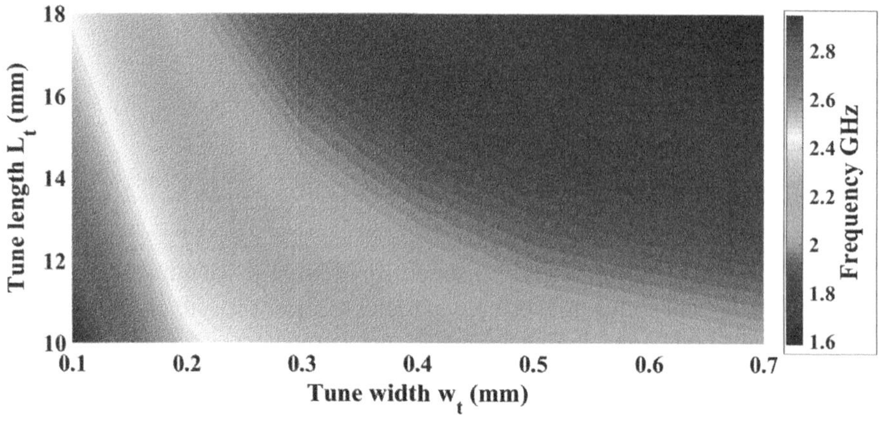

Figure 4.20. *Resonance frequency calculated using [4.1]–[4.3] for reset (high impedance) state of CBRAM/MIM switch for electronically reconfigurable band-stop filter design N2. For a color version of this figure, see www.iste.co.uk/jayakrishnan/technology.zip*

The maximum calculated error for set (low impedance) and reset (high impedance) states, using the mathematical model, are, respectively, 30 MHz and 110 MHz. Comparisons of resonance frequency values obtained using the mathematical model, and in experiments, for the presented device N2 (with $L_t = L_3 = 14.7$ mm, and $w_t = w_4 = 0.3$ mm) are shown in Table 4.3.

This design technique using the given electrical model and design equation could be used for a quick adaptation of the proposed filter for a specific need. The adapted model could then be optimized in simulation, to exactly match the required needs, and realized in fabrication.

One should also note that the set (low resistance) state resistance value of MIM switch filament resistance has an attenuating effect on the stop band isolation of the filter. The lower the MIM filament resistance R_{MIM}, the higher the stop band isolation in set state. This is similar to the phenomenon explained in Chapters 2 and 3, respectively, for the RF switch and chipless RFID resonator, in the same context. This analysis is omitted from this section to avoid redundancy.

CBRAM/MIM switch state	Experimental realization *Fr* (GHz)	Mathematical model using equations [4.1]–[4.3] *Fr* (GHz)	Error observed among experiment and mathematical model (MHz)
Set (low impedance) state	1.421	1.432	11
Reset (high impedance) state	2.010	2.026	16

Table 4.3. *Comparison of experimentally obtained and mathematical model values of resonance frequency for design N2*

4.2.2. *Discussion of extension of the proposed idea of CBRAM/MIM RF switching to more efficient filter topologies*

The above-proposed filter designs are basically first order and are limited in terms of quality of stop band isolation and bandwidth. Nevertheless the presented experiment is a proof of concept and highlights the advantages and feasibility of using proposed integrated CBRAM/MIM RF switch technology for electronically reconfigurable filter applications. All of the fabricated devices were tested for more than 50 cycles of operations and were observed to be still functional.

In Chapters 2 and 3, we demonstrated the realization of passive RF devices with integrated CBRAM/MIM switches on paper substrates. This shows that the presented design could also be realized similarly, without any intrusive soldered electrical components, for low-cost and flexible RF electronics applications, on substrates like paper.

The proposed design practice is an assuring candidate for robust low-power, low-cost and low-weight applications, for example, in Internet of Things (IoT)-based stand-alone systems. This technique could also be protracted to all types of planar filters. These could be optimized for specific and desired filtering characteristics, either using well-known classic design techniques (Makimoto and Yamashita 2001; Pozar 2005), or using assorted contemporary topologies outlined in various studies (Chaudhary *et al.* 2012; Chen *et al.* 2017).

In the following, we discuss some strategies and techniques for extending the applications of the proposed idea of electronically reconfigurable filters using non-volatile CBRAM switches for more efficient and versatile performance.

4.2.2.1. *Example of an electronically reconfigurable band-pass filter*

In this section, an extension of the proposed CBRAM/MIM switch technology for RF filtering applications to electronically reconfigurable band-pass filters is presented in brief with the help of full-wave electromagnetic simulation studies.

Figure 4.21 show the topology of a reconfigurable band-pass filter with an integrated MIM switch. The presented band-pass filter is based on a resonator element similar to a half wavelength resonator, with an MIM switch integrated into the structure by tapering the trace width of the resonator element. This is similar to a coupled line band-pass filter using a modified hairpin resonator element (Chaudhary *et al.* 2013). RF power is fed to the resonator through proximity coupled microstrip transmission lines.

Here, simulation is carried out by modeling the geometry of the filter, as shown in Figure 4.21, and operating the MIM switch using a lumped resistor with values 1 Ω and 1 MΩ, respectively, for set (low impedance) and reset (high impedance) states, as described for the simulation of band-stop filters above. Here, the MIM switch is integrated among the trace widths marked "W4" (copper–active electrode) and "W5" (aluminum–inert electrode) in Figure 4.21 by sandwiching 600 nm of nafion (ion conductor). Figure 4.22 shows the simulated S-parameters of the electronically reconfigurable band-pass filter for set (low impedance) and reset (high impedance) states of the switch.

In Figure 4.22, we can see that the filter shows a pass band centered at 1.5073 and 2.5241 GHz, respectively, for set (low impedance) and reset (high impedance) states of the integrated MIM switch.

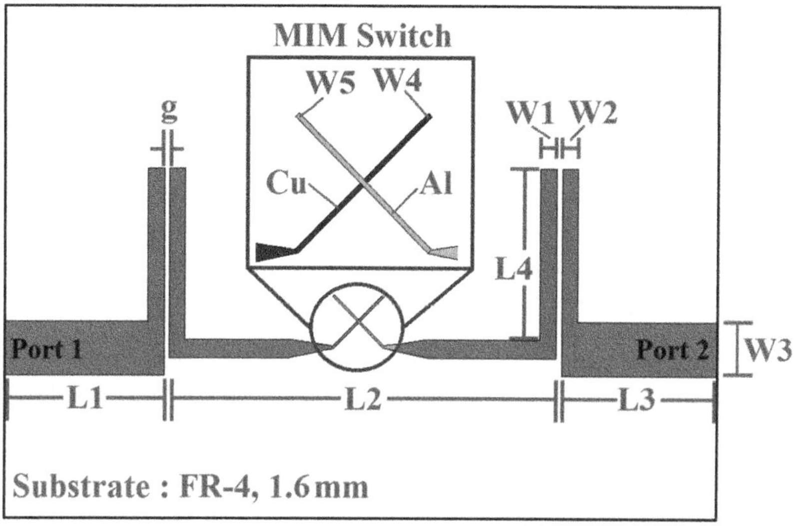

Figure 4.21. *Topology of proposed electronically reconfigurable band-pass filter (L1 = 11 mm, L2 = 27.8 mm, L3 = 11 mm, L4 = 9.5 mm, W1 = W2 = 1 mm, W3 = 3 mm, W4 = W5 = 0.1 mm, g = 0.1 mm). The CBRAM/MIM switch is formed by sandwiching a 600 nm layer of nafion (ion conductor) among the traces marked by width "W4" (copper) and "W5" (aluminum)*

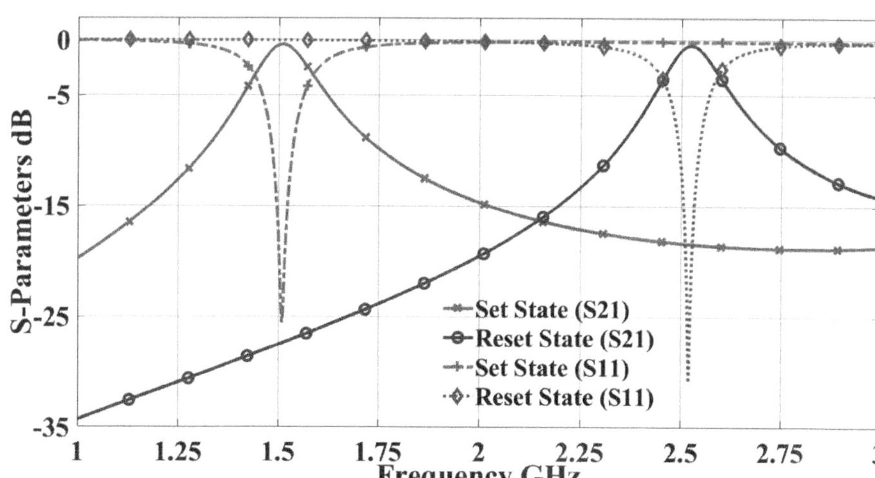

Figure 4.22. *Simulated (full-wave) RF response of electronically reconfigurable band-pass filter. For a color version of this figure, see www.iste.co.uk/jayakrishnan/technology.zip*

In the passband center frequency, the device shows an insertion loss less than –0.5 dB and isolation greater than –18 dB subsequently for both set and reset states of the integrated MIM switch. The –3 dB bandwidth is calculated to be greater than 140 MHz for both pass bands of the filter. This example serves as a proof of extension of the proposed MIM switch technology for band-pass filter applications. Displacement among the two pass bands of the filter at set and reset states could be regulated by controlling reset (high impedance) state capacitance of integrated CBRAM/MIM switch, as explained above for the electronically reconfigurable band-stop filter.

4.2.2.2. Techniques to improve performance characteristics of presented design of electronically reconfigurable band-stop filters

Filters N1 and N2 presented herewith are based on an elementary design using a simple operating principle, and are limited in performance factors like filtering bandwidth, stop band isolation and roll-off rate (order). The major goal of this realization is not in achieving smooth filter performances, but rather the proof of concept of using CBRAM switches for electronically reconfigurable RF and microwave filter applications. In addition, we demonstrate the integration of these switches in general purpose RF and microwave device designs, and fabrication of these devices using a simple manufacturing technique suitable for mass production, without using any "clean room" techniques, as explained in detail in Chapter 2.

Nevertheless, the efficiency of presented designs N1 and N2 in the above context of performance characteristics of filtering action could be enhanced in several ways. One simple technique involves introducing multiple identical resonant stub sections with integrated CBRAM/MIM switches, which could improve the bandwidth of operation, stop band isolation and roll-off rate (order). This concept is demonstrated with the help of full-wave electromagnetic simulations in the following.

A filter topology similar to N1, with a single stub, optimized for operation at 2.5 GHz, shows a stop band isolation of –10.5 dB with 80 MHz of (–10 dB) bandwidth, and a (–3 dB) roll-off rate of 5.8 dB/GHz, in simulation. This topology, upon the addition of one and then two identical stubs, the second similar to the first stub with an integrated MIM switch, shows an enhancement as shown in Table 4.4. The topology of the simulated filter design is shown in Figure 4.23, and Figure 4.24 shows the obtained S21 response. In this simulation, MIM switches in all identical stubs present were simultaneously switched to set (low impedance) and reset (high impedance) states, respectively, for a given set or reset state of the filter.

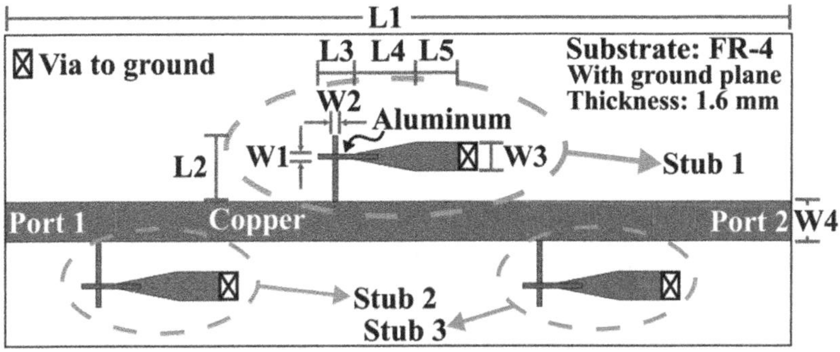

Figure 4.23. *Topology of electronically reconfigurable band-stop filter with multiple (identical) resonator stubs (L1 = 60 mm, L2 = 5 mm, L3 = 2.5 mm, L4 = 5 mm, L5 = 3.85 mm, W1 = W2 = 0.1 mm, W3 = 2 mm, W4 = 3 mm). For a color version of this figure, see www.iste.co.uk/jayakrishnan/technology.zip*

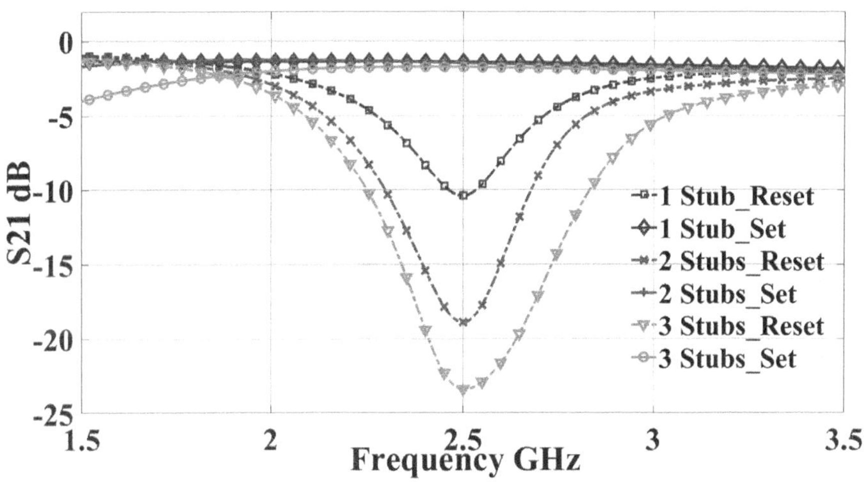

Figure 4.24. *Simulated (full-wave) RF response of electronically reconfigurable band-stop filter for different numbers of (identical) resonator stubs. For a color version of this figure, see www.iste.co.uk/jayakrishnan/technology.zip*

One other limitation of the presented filter topologies is that they are limited to single frequency tuning for design N1, and dual frequency tuning for design N2, in comparison to applications utilizing classic RF switch technologies such as PIN diodes (Chaudhary *et al.* 2012; Chen *et al.* 2017) or field effect transistors, which show a continuous tuning functionality in a desired bandwidth. However, "non-volatility" obtained from proposed CBRAM/MIM switches is the major point of interest and highlighted innovation of presented work. This cannot be implemented with MEMS or classic solid-state switches like PIN diodes. Moreover, in this design, the CBRAM/MIM switches are directly integrated into the topology, without any techniques like soldering or conductive gluing, which is not the case for above-mentioned classic applications in general.

Number of stubs	Isolation at center frequency of stop band	Bandwidth (–10 dB)	Roll-off rate (–3 dB)
1	–10.5 dB	80 MHz	5.8 dB/GHz
2	–18.92 dB	380 MHz	7.6 dB/GHz
3	–23.41 dB	580 MHz	8.4 dB/GHz

Table 4.4. *Enhancement in filter parameters as a function of number of identical resonator stubs*

Anyhow, multiple frequencies tuning in presented filters, similar to design N2, could be achieved by using multiple integrated CBRAM/MIM switches in the resonant stub to tune its electrical length. A full-wave electromagnetic simulation using design N2 was undertaken to present this concept, in which the resonant stub was segmented by the addition of multiple MIM switches. This was done by cascading two 5 mm long stub segments in series, at the open circuit end of the resonant stub, using the electrical model of integrated MIM switches connecting the segments, as shown in Figure 4.25. Here, we use an RC combination of lumped capacitor C_{MIM} and resistor R_{MIM} to model the MIM switch. The value of C_{MIM} is designated to be 1.4 pF, and R_{MIM} is chosen to be 1 and 1 MΩ, respectively, for set (low impedance) and reset (high impedance) states of the MIM switch. These values are chosen based on previous realizations and the electrical model presented in Table 4.2. Results of full-wave electromagnetic simulation of this model are shown in Figure 4.26.

Electrical models of the MIM switch are used here instead of a real-time model for simplicity and fast simulation; for a desired practical application, these could be replaced with a typical physical prototype of integrated MIM switches with appropriate dimensions. Here, as expected, the simulation response of this design shows that the center frequency of the stop band could drift along a range of 900 MHz from the lowest center frequency (obtained typically when all switches are short circuited or in set state), as shown in Figure 4.26, through eight frequency points along the bandwidth for different combinations of MIM switches. The precision, range and number of frequency positions in tuning could be optimized as a function of the number of switches, and by controlling the reset (high impedance) state capacitance C_{MIM} of the switches as explained in section 4.2.1.3.

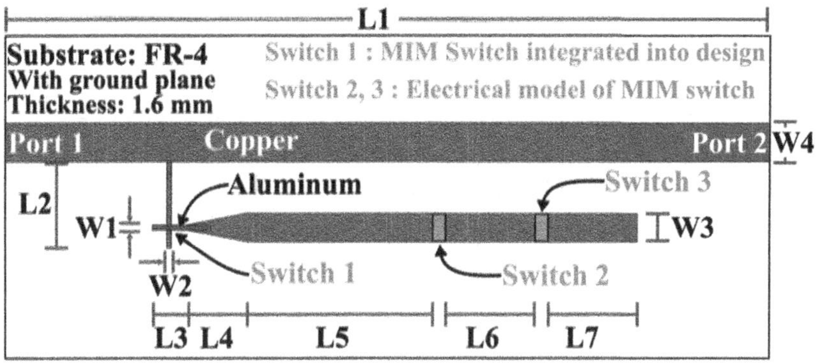

Figure 4.25. *Topology of proposed model of band-stop filter with multifrequency tuning. L1 = 60 mm, L2 = 5 mm, L3 = 2.5 mm, L4 = 5 mm, L5 = 14.7 mm, L6 = L7 = 5 mm, W1 = 0.3 mm, W2 = 0.1 mm, W3 = 2 mm and W4 = 3 mm. For a color version of this figure, see www.iste.co.uk/jayakrishnan/technology.zip*

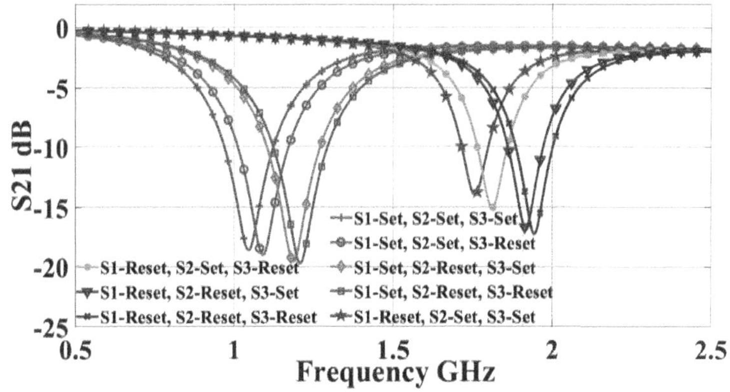

Figure 4.26. *Simulated (full-wave) RF response of electronically reconfigurable band-stop filter with multifrequency tuning presented in Figure 4.25. For a color version of this figure, see www.iste.co.uk/jayakrishnan/technology.zip*

4.3. MIM switches for electronically pattern reconfigurable antenna applications

Electronic pattern agility or reconfigurability of antennas is a well-investigated field in the current decade. This concept could significantly enhance the performance and efficiency of a communication link. A good number of congenial designs have been reported recently in the literature, giving priority to multidirection beam steering or beam forming, for IoT and fifth generation cellular network (5G) related applications (Lim and Ling 2007; Ding and Wang 2013; Farzami *et al.* 2017; Lin *et al.* 2017; Lu *et al.* 2017; Chen *et al.* 2018). However, in most of these cases, electronic reconfigurability is attained utilizing active and volatile switching devices such as solid-state semiconductor or MEMS switches. Such switches require a continuous power supply to maintain their impedance state, despite their excellent results, which critically impacts the power budget, and restricts their use in low-power and passive stand-alone devices. Furthermore, these switches are usually available in the form of modular integrated circuits, and are generally not very suitable for emerging trends in printed electronics, needing to be added into a target design by intrusive methods like soldering or conductive gluing.

The recent discovery of memristive devices and their application as non-volatile RF switches, such as CBRAM/MIM switches (Jayakrishnan *et al.* 2017; Pi *et al.* 2015; Kozicki, West 1998; Nessel, Lee 2011), give a ray of hope for a new revolution of passive microwave devices, integrated with non-volatile switching devices.

A handful of scientific publications, some of them from a decade ago, predicted the possibility of using memristive switches for passive RF and microwave devices, some especially for reconfigurable antenna applications (Bray and Werner 2010; Wang *et al.* 2011; Gregory and Werner 2015; Dragoman *et al.* 2017). Yet, one of the limitations of these reported devices is their sophisticated and complex structure, and the requirement of a delicate “clean room” facility for the fabrication processes. However, in this book, we try to address and mitigate this limitation of memristive switches, through a strategy combining smart materials and simple manufacturing techniques, for fabrication of non-volatile CBRAM RF switches.

In the following sections, we present the design, fabrication and experimental characterization of an electronically pattern steerable antenna, utilizing integrated polymer ion conductor-based CBRAM/MIM RF switches. This is accomplished through the techniques for realization of CBRAM-based RF switches on rigid substrates explained in Chapter 2, without involving any clean room process for fabrication. This work was reported for the first time in Jayakrishnan *et al.* (2019a).

4.3.1. *Electronically radiation pattern steerable antenna using CBRAM/MIM RF switches (design and fabrication)*

In this section, we present the design and realization of an electronically pattern steerable half wave dipole antenna using integrated non-volatile CBRAM/MIM switch technology. The topology of the electronically pattern reconfigurable antenna using an integrated CBRAM/MIM switch is shown in Figure 4.27, and its dimensions are depicted in Table 4.5. This design geometry is adapted from a parasitic element (PE)-loaded classic dipole antenna model. Radiation pattern steering of this antenna is achieved by loading an MIM cell into the PE to tune its electrical length. This PE is placed in parallel to the dipole length, and the MIM switch is integrated at its center, parallel to the feed center of dipole radiator. Figure 4.27 shows this PE, with MIM switch integrated into it, by interposing a 600 nm layer of nafion (ion-conductor) among the trace width marked as "W6" (copper – active electrode) and "W7" (aluminum – inert electrode). The geometry of the PE is adapted by tapering its width at the center to reduce effective capacitance due to overlapping sections, which forms the active and inert electrode of the MIM switch.

A classic and well-known microstrip to slot-line transition is used as a balanced-to-unbalanced signal converter or "balun" (Li *et al.* 2009) to feed the presented dipole antenna. Insertion loss (or S11) of the antenna is optimized in the full-wave electromagnetic simulation using CST Microwave Studio by adjusting the geometric parameters L3, L4, L5, D, S and g. Filament formation (switching mechanism) of the integrated MIM switch is modeled and operated in simulation studies, using a lumped resistor, with values similar to the experimentally obtained filament resistance of a typical MIM cell, as explained in Chapter 2, for fast and easy computation.

The working mechanism of the presented electronically pattern steerable antenna model can be explained as follows. Here, for set (ON) and reset (OFF) states of the integrated MIM switch, the PE respectively acts as a reflector and as a director element. The CBRAM/MIM switch connected to the PE as shown in Figure 4.27 could be modeled as an RC parallel network, where the filament resistance R_{MIM} is in parallel to the capacitance C_{MIM} due to the geometry of the switch, as explained in Chapter 2.

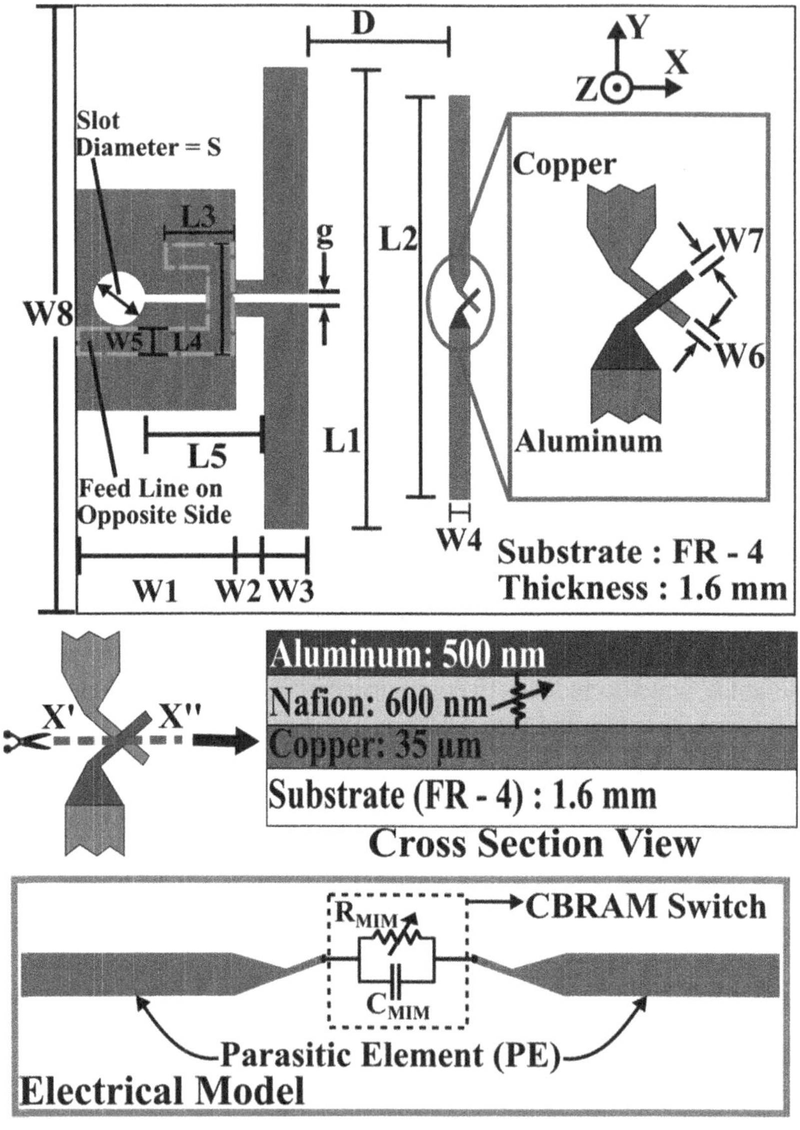

Figure 4.27. *Topology of electronically pattern reconfigurable antenna with integrated MIM switch and its layer structure. Inset in green border shows the electrical model of the MIM switch connected to the parasitic element. For a color version of this figure, see www.iste.co.uk/jayakrishnan/technology.zip*

Parameters	L1	L2	L3	L4	L5	L6
Dimension (mm)	50	44	8	12	13	80
Parameters	W1	W2	W3	W4	W5	W6
Dimension (mm)	18	3	5	2	3	0.1
Parameters	W7	W8	S	D	g	–
Dimension (mm)	0.1	60	6	16	0.5	–

Table 4.5. *Dimensions of electronically pattern reconfigurable antenna geometry shown in Figure 4.27*

The filament resistance R_{MIM} is of a very low value (of the order of a few Ω), during the set (ON) state of the switch, and hence short circuits C_{MIM}, neglecting it from the current path. Thus, it allows the PE to exhibit its natural electrical length (similarly to when the switch is absent) and consequently act as a reflector, like it was chosen to do. On the other hand, in the reset (OFF) state of MIM switch, R_{MIM} is a large value (of the order of a few kΩ to MΩ) similar to an open circuit, and hence the major current path is relocated through C_{MIM}. This results in reducing the effective electrical length of the PE as C_{MIM} is added in series to its equivalent RLC model, hence forcing it to act as a director element.

Simulated (full-wave) return loss characteristic of the presented electronically pattern reconfigurable antenna model is shown in Figure 4.28. Radiation patterns (gain in dBi) of this antenna obtained through full-wave electromagnetic simulation for a common frequency with adequate return loss matching (2.23 GHz) are depicted in Figures 4.29 and 4.30, respectively, for H-plane and E-plane. One could see from these radiation patterns that the presented electronically pattern reconfigurable antenna exhibits the desired pattern steering performances for set (ON) and reset (OFF) states of integrated CBRAM/MIM switches. The radiation patterns of this antenna are steered for an angle of ±180°, respectively, for the set (ON) and reset (OFF) states of integrated CBRAM switches both in E and H planes. The electronic pattern steering action can be understood more clearly and explicitly by looking at the simulated (full-wave) 3D radiation pattern shown in Figures 4.31 and 4.32, which show a beam maxima steered to +X direction and –X direction, correspondingly for set (ON) state and reset (OFF) states of integrated MIM switch. Further discussions on performance factors of the presented antenna are given in detail in section 4.3.1.1 along with experimentally obtained results. The surface current patterns on the antenna and PE for set (ON) and reset (OFF) states of the integrated MIM switch are shown in Figures 4.33 and 4.34.

The primary target of this experiment was to prove the feasibility of utilizing CBRAM/MIM switches for reconfigurable antenna applications, which is accomplished with affirmative outcomes. A detailed analysis into the mathematical level of the modeling of the antenna performance is beyond the scope of this book and hence is omitted to maintain brevity. However, a general explanation on the working mechanism is given below for the reader's general understanding.

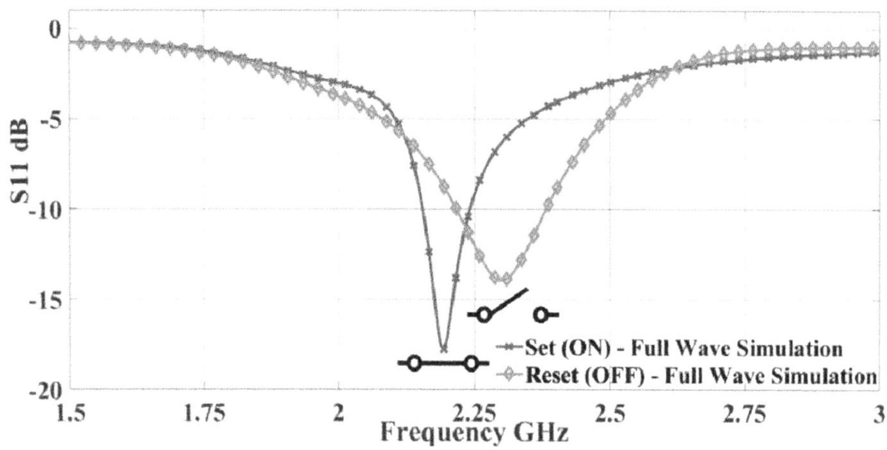

Figure 4.28. *Simulated (full-wave) return loss (S11) characteristics of electronically pattern reconfigurable antenna*

The presented antenna is similar in design and working principle to a classic and well-known Yagi Uda antenna (Balanis 2005), although with enhanced electronic reconfigurability features. Here, the physical size of the PE is chosen to be less than that of the dipole, and is optimized in simulation to exhibit a higher electrical length, by introducing the taper sections. In this way the fundamental resonance frequency of the resonator section will be lower than that of the dipole and acts as a reflector element. Then an MIM switch is integrated at the center of the PE as shown in Figure 4.27. In the set (ON) state of the switch (which is similar to a short circuit in its center), the PE still acts as a reflector. Inversely in the reset (OFF) state of the MIM switch, the PE could be modeled as two separate dissected sections connected together with a capacitor section (the resistor in parallel to the capacitor is a very high value in the reset state of the MIM switch and can be ignored by current path), as shown in the inset of Figure 4.27. This in turn reduces the effective electrical length of the PE, forcing it to act as a director element, in association with the dipole. In this case, the fundamental resonance frequency of this PE would be higher than that of the dipole.

This observation can also be justified by the radiation phase behavior of two coupled resonators. It is well known that there will be a phase inversion of 180° in the radiation pattern of a resonator just before and after the resonance. Thus, we can conclude that in the set (ON) state of the MIM switch, the PE resonance is lower than that of the dipole, and it exhibits destructive interference, resulting in the radiation pattern pointing in the –X direction in Figure 4.27. Inversely in the reset state of the MIM switch, the resonance of PE is higher than that of dipole and hence there is a 180° shift in phase, which results in constructive interference of the radiation patterns, in which the net pattern points in the +X direction.

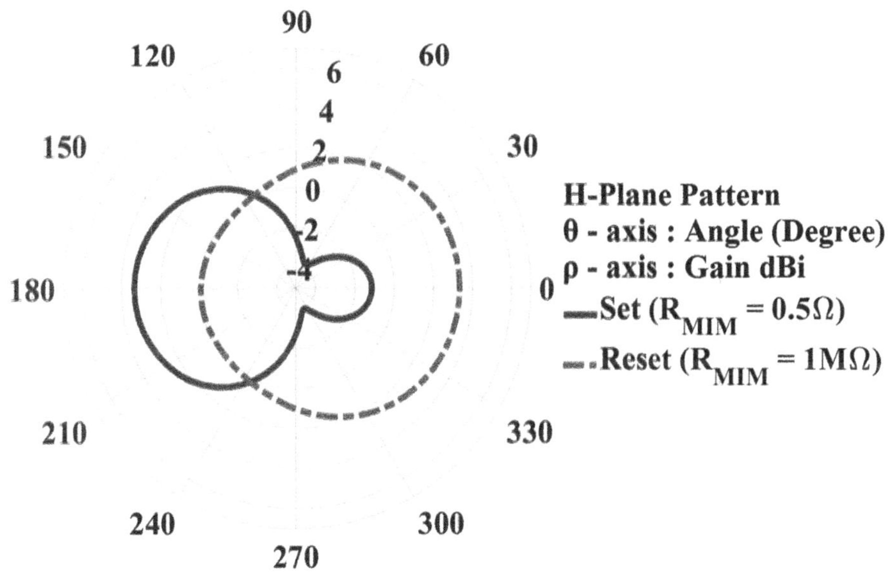

Figure 4.29. *Simulated (full-wave) H-plane radiation pattern of electronically pattern reconfigurable antenna*

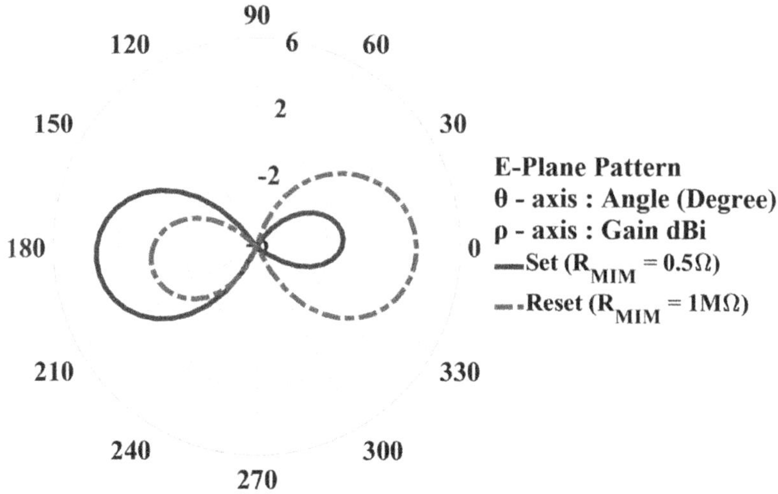

Figure 4.30. *Simulated (full-wave) E-plane radiation pattern of electronically pattern reconfigurable antenna*

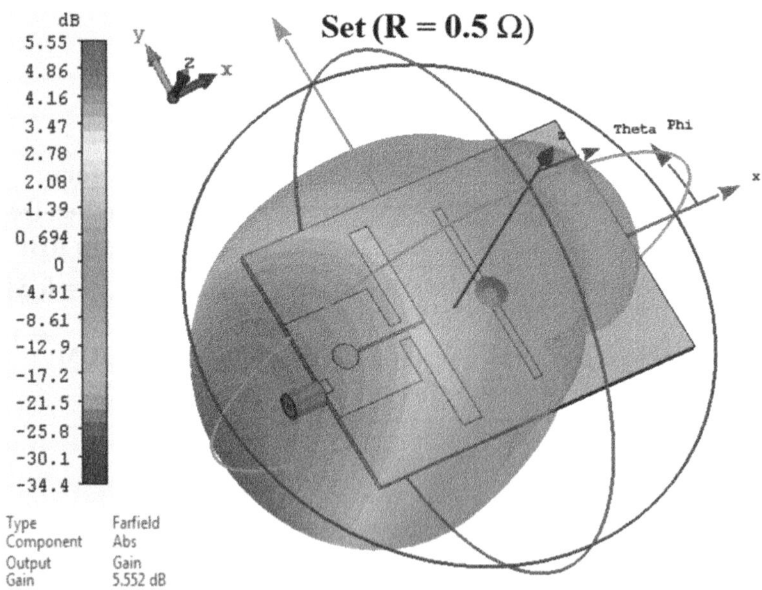

Figure 4.31. *Simulated (full-wave) 3D radiation pattern of electronically pattern reconfigurable antenna for set (ON) state of integrated MIM switch. For a color version of this figure, see www.iste.co.uk/jayakrishnan/technology.zip*

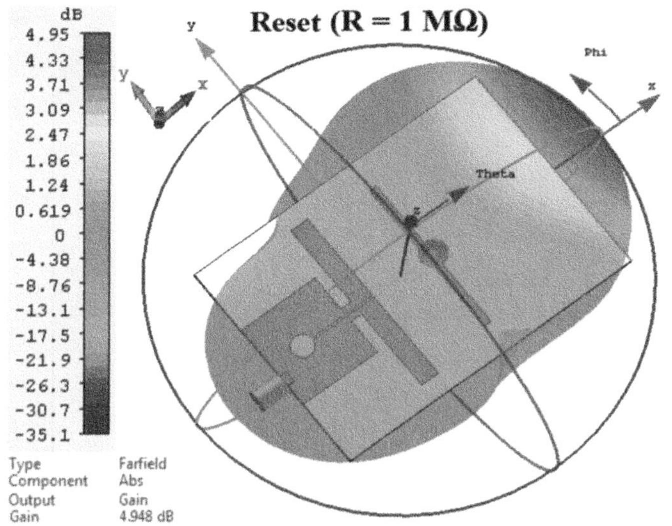

Figure 4.32. *Simulated (full-wave) 3D radiation pattern of electronically pattern reconfigurable antenna for reset (OFF) state of integrated MIM switch. For a color version of this figure, see www.iste.co.uk/jayakrishnan/technology.zip*

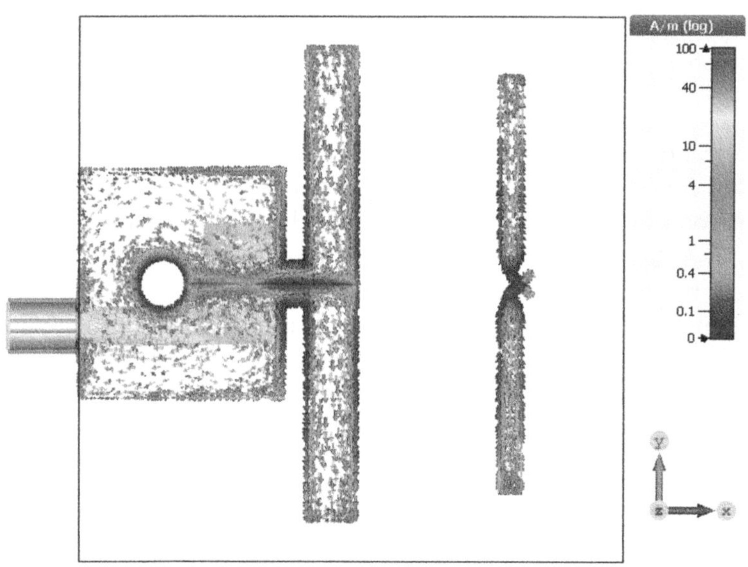

Figure 4.33. *Simulated (full-wave) surface current patterns on the antenna and parasitic element (which acts as a reflector in this case), at resonance frequency of dipole, for set state of MIM switch. For a color version of this figure, see www.iste.co.uk/jayakrishnan/technology.zip*

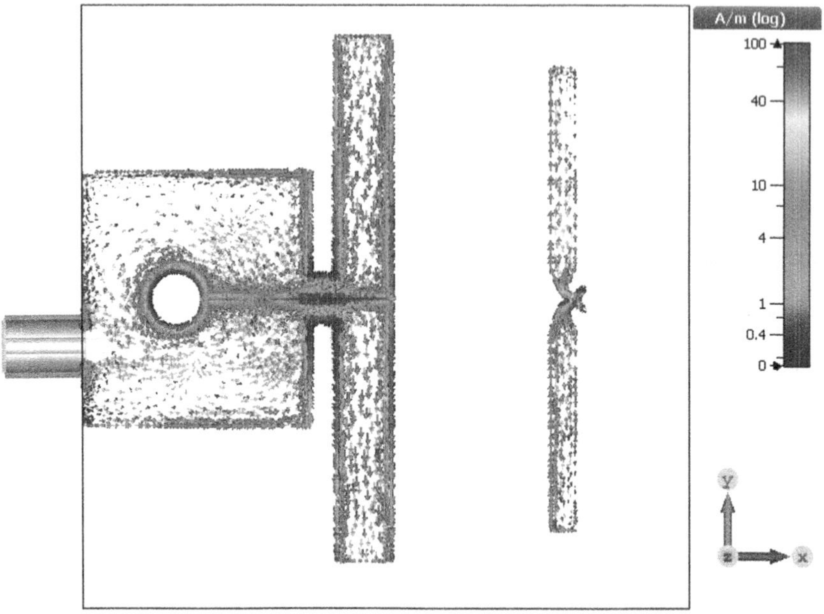

Figure 4.34. *Simulated (full-wave) surface current patterns on the antenna and parasitic element (which acts as a director in this case), at resonance frequency of dipole, for reset state of MIM switch. For a color version of this figure, see www.iste.co.uk/jayakrishnan/technology.zip*

Fabrication of the proposed electronically pattern reconfigurable antenna is undertaken using techniques explained in Chapter 2 for CBRAM RF switches on rigid substrates, on a classic double side copper cladded 1.6-mm thick FR-4 laminate. Patterned copper on the copper cladded substrate (marked by width "w6" in Figure 4.27) and thermal vapor deposited aluminum (marked by width "w7" in Figure 4.27), respectively, are the active and inert electrodes. Spin-coated nafion with a layer thickness of 600 nm serves as the ion conductor layer. Figure 4.35 shows the photograph of the fabricated reconfigurable antenna, with a zoom image of the MIM switch area in the inset.

The MIM switch is operated using DC pulses, generated from a Keithley 2400 source meter with the help of a custom assembled Labview-GUI as explained in Chapter 2, and fed through metallic parts of the PE carrying the integrated MIM switch, using ordinary crocodile clips. The switch is set (ON) using a triangular DC pulse of peak voltage 16 V with compliance current limit of 500 μA and reset (OFF) using a square pulse of peak voltage –20 V with a compliance current limit of 150 mA. Set (ON) state resistance of the MIM switch is observed to be among 15–20 Ω, and reset (OFF) state is in the range of 100–200 MΩ in this experiment.

Figure 4.35. *Photograph of fabricated electronically pattern reconfigurable antenna (inset shows zoomed microphotograph of MIM switch area). For a color version of this figure, see www.iste.co.uk/jayakrishnan/technology.zip*

4.3.1.1. *Experiments, results and discussion*

Radiation pattern measurements of the fabricated antenna are carried out using the MVG Starlab® 3D radiation pattern measurement system (MVG 2014). Starlab is a fully automatic radiation pattern measurement system, which is capable of capturing real-time 3D gain and directivity patterns. Starlab is operated on our experimental platform with the support of an Agilent N5222A network analyzer and a control computer.

Figure 4.36 shows the photograph of a functional Starlab measurement system. Calibration of this measurement system is made by capturing the radiation pattern of a known broadband horn antenna and quantifying this captured pattern with a predetermined gain-angle-frequency data file of the same antenna. Starlab uses a patented interlaced scanning feature (similar to synthetic aperture radar scanning) using a near-field probe technology to capture the near-field radiation pattern of the antenna under test, which is later converted automatically to a farfield pattern using associated data processing software.

Figure 4.36. *MVG Starlab® automatic 3D radiation pattern measurement system. For a color version of this figure, see www.iste.co.uk/jayakrishnan/technology.zip*

Experimentally obtained return loss characteristics of the fabricated electronically pattern reconfigurable antenna for set (ON) and reset (OFF) states of the integrated MIM switch, along with full-wave simulation results for comparison, are shown in Figure 4.37. The return loss (S11) response shows a –10 dB bandwidth of 54 MHz centered at 2.23 GHz in common for both set (ON) and reset (OFF) states of the MIM switch, which are the possible configuration modes of the antenna. A slight difference in magnitude of return loss characteristics between simulation and experiment is accounted for by the small differences between simulated and realized models due to fabrication tolerances.

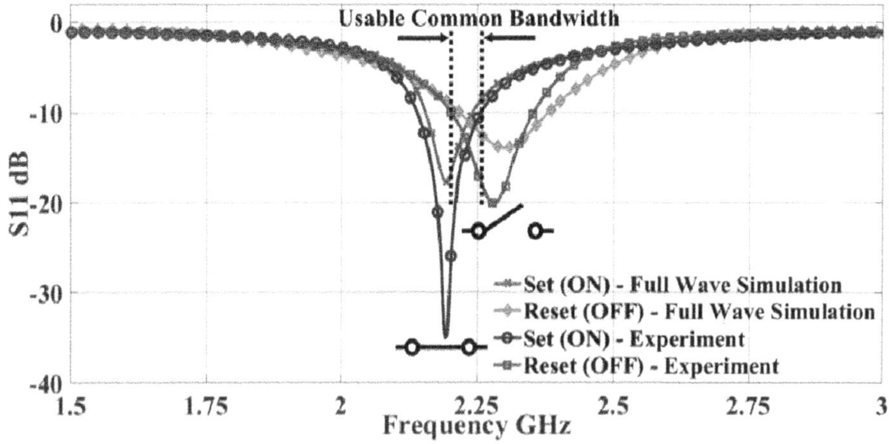

Figure 4.37. *Experimental and simulated (full-wave) return loss (S11) characteristics of electronically pattern reconfigurable antenna. For a color version of this figure, see www.iste.co.uk/jayakrishnan/technology.zip*

Experimentally obtained H-plane and E-plane radiation (gain in dBi) patterns at 2.23 GHz are presented in Figures 4.38 and 4.39, respectively. An electronic steering of ±180° is attained for the major lobe of the radiation pattern, for set (ON) and reset (OFF) states of the switch, in the plane perpendicular (along the X-direction in Figure 4.27) to the axis of the dipole along its length. A beam maxima toward the positive X direction, with an antenna gain magnitude of 3.825 dBi and a back-lobe magnitude of –1.596 dBi, is obtained in the H-plane for the reset (OFF) state of MIM switch. This pattern is steered by 180°, along the negative X direction in the set (ON) state of the MIM switch, to exhibit maxima of 3.051 dBi for the front lobe, with a back-lobe magnitude of –2.110 dBi (see Figure 4.38). Similarly for E-plane pattern we can observe, along the positive X direction, a magnitude 4.412 dBi for the major lobe, with a back-lobe magnitude of –1.511 dBi for the reset (OFF) state. For the set (ON) state, we can see a 180° steering along the negative X direction to exhibit maxima of 3.051 dBi and a back-lobe magnitude of –0.939 dBi (see Figure 4.39).

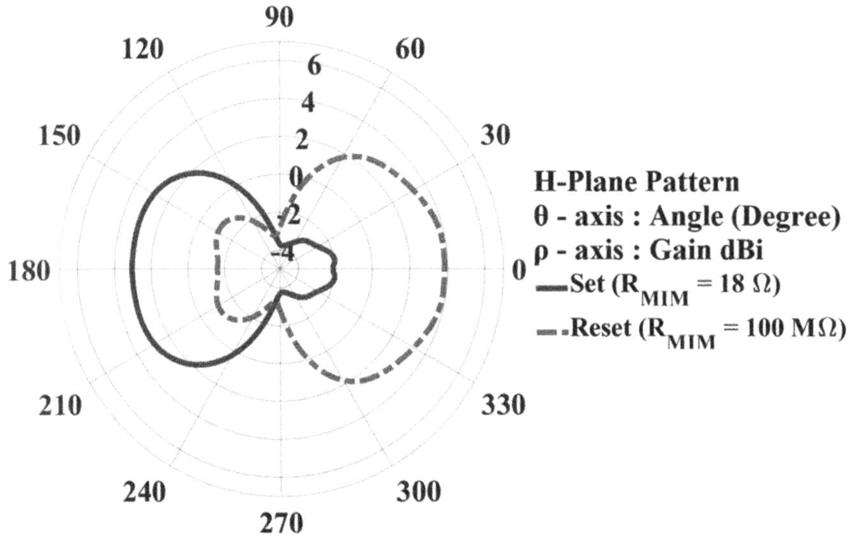

Figure 4.38. *Experimentally obtained H-plane radiation pattern of electronically pattern reconfigurable antenna*

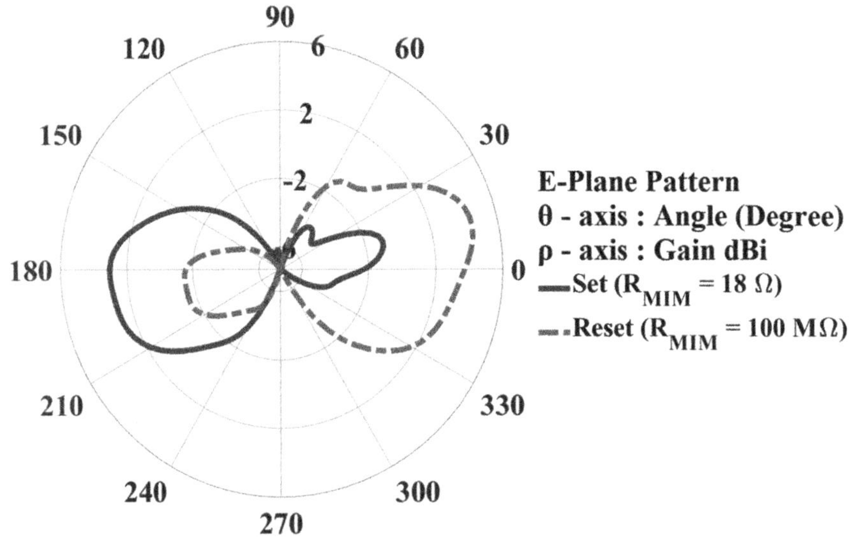

Figure 4.39. *Experimentally obtained E-plane radiation pattern of electronically pattern reconfigurable antenna*

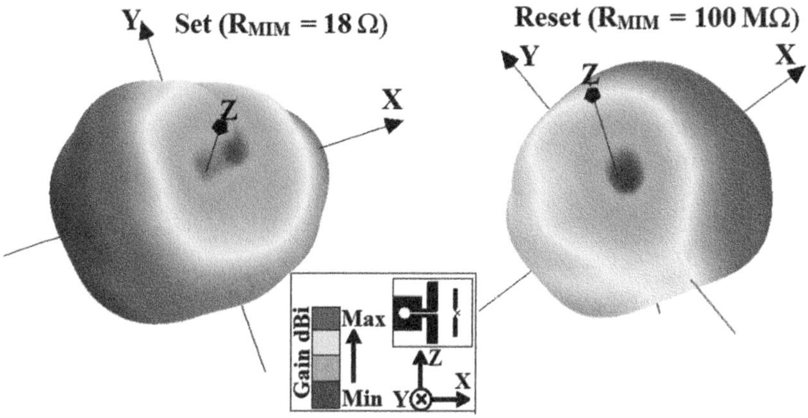

Figure 4.40. *Experimentally obtained 3D radiation pattern of electronically pattern reconfigurable antenna. For a color version of this figure, see www.iste.co.uk/jayakrishnan/technology.zip*

In summary, an electronic steering of ±180° from positive to negative end-fire directions of the antenna is achieved in the 3D radiation pattern by switching the integrated MIM cell. Figure 4.40 presents experimentally obtained 3D radiation patter (gain) of the antenna for set (ON) and reset (OFF) states of the integrated MIM switch. Placement of antenna between coordinate axes for the radiation pattern in Figure 4.40 is given in an inset. This representation validates the above explanation of the 180° pattern steering expected from the antenna, along the plane perpendicular to the axis of the dipole along its length. In the whole of the experiment, the antenna shows a maximum gain of 4.4, and 3.051 dBi, respectively, in reset (OFF) and set (ON) states of MIM switch, along the beam maxima.

Slight differences in gain levels between the experimentally obtained and simulated radiation patterns in the reset state of the integrated MIM switch are accounted for by difference in losses associated with the substrate, and thermal vapor deposited metals used in real fabrication, and in simulated model. One should especially note that the thickness of the inert electrode of the MIM switch (~500 nm), using aluminum (see Figure 4.27), is well below the skin depth at operating frequency (which is ~1.5 μm for bulk aluminum at 3 GHz). This would contribute to reducing the radiation efficiency of the antenna, however with the benefit of an added functionality – non-volatile electronic reconfigurability. In the set (ON) state of the switch, we can see that there is even more reduction of gain for the experimental results when compared to the simulated outcome. This could be explained, in addition to the above reason, as a reduction in radiation efficiency (or gain) of the PE, with the set (ON) state impedance of MIM switch. A detailed discussion of this phenomenon and possible productive utilization of the same is provided in the following section.

4.3.1.2. *Effect of filament resistance of integrated CBRAM/MIM switch on antenna gain*

As explained in Chapter 2, the set (ON) state filament resistance (i.e. impedance typically lower than a few hundred of ohms) of a CBRAM/MIM switch has a notable controlled attenuation effect on the RF energy passing through it. In the case of the presented electronically pattern reconfigurable antenna, this phenomenon is seen to be a control parameter for antenna gain in the set (ON) state of the integrated MIM switch. Figures 4.41 and 4.42 show, respectively, the plots of H-plane and E-plane radiation patterns of the antenna (shown in Figure 4.27), showing variation in gain as a function of the set (ON) state resistance of integrated MIM switch. These plots are obtained through full-wave electromagnetic simulation in CST Microwave Studio.

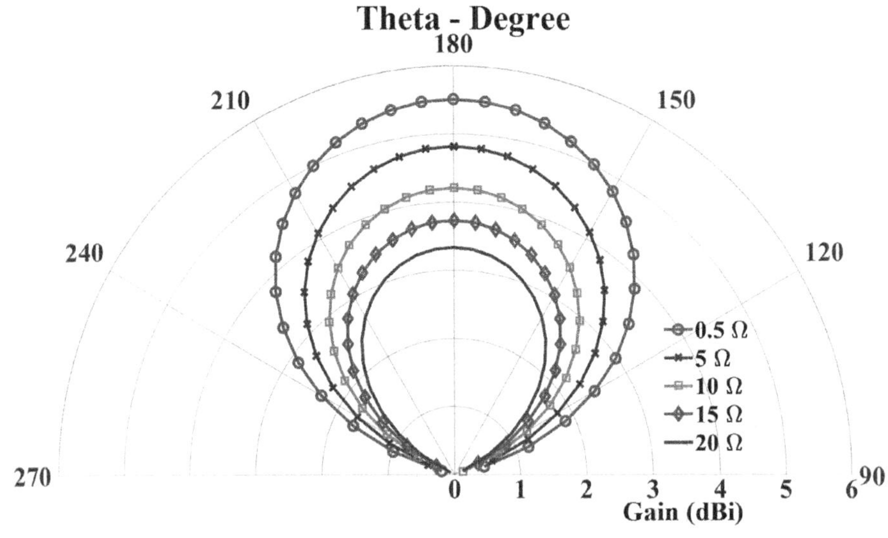

Figure 4.41. *Variation of H-plane gain (full-wave simulation) of electronically pattern reconfigurable antenna as a function of set (low impedance) or ON state resistance of integrated MIM switch. For a color version of this figure, see www.iste.co.uk/jayakrishnan/technology.zip*

From these plots, we can see that the set (ON) state resistance of the integrated MIM switch in the antenna could be used as a control parameter for the gain of the antenna in its respective direction in this proposed design. This is believed to be a function of the radiation efficiency of the antenna contributed by the PE, which acts as a reflector in the set (ON) state of the switch. Similar variation is also observed in the back-lobe of the antenna, with resistance variation. The back-lobe magnitude is seen to increase with increases in set (ON) state resistance. However, the maximum gain value is seen to be less than –1 dBi in this case, and it is excluded from these plots to avoid crowding.

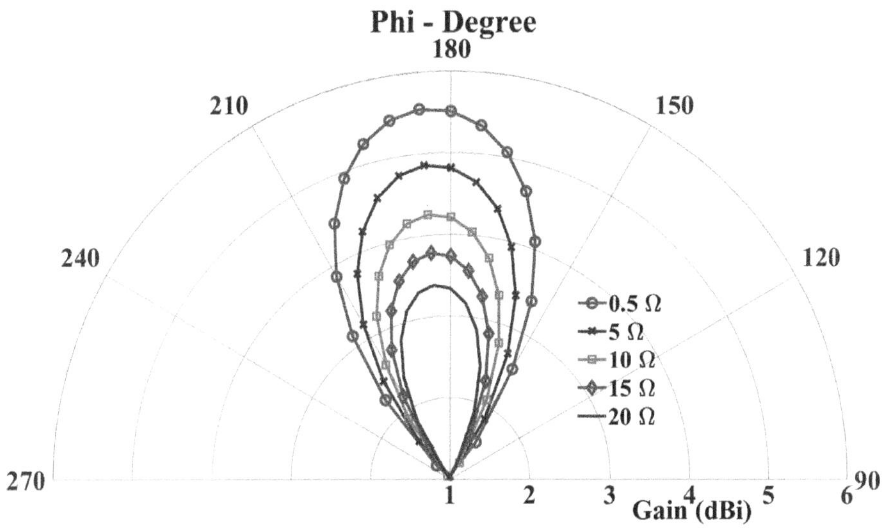

Figure 4.42. *Variation of E-plane gain (full-wave simulation) of electronically pattern reconfigurable antenna as a function of set (low impedance) or ON state resistance of MIM switch. For a color version of this figure, see www.iste.co.uk/jayakrishnan/technology.zip*

As mentioned previously in the case of the CPW RF switch described in Chapter 2, an efficient technique to control the filament resistance of the MIM switch could help in the development of a gain reconfigurable antenna. Such a gain reconfigurable antenna could be a good candidate for applications such as secure communications in a given range, limiting network access like Wi-Fi or mobile communication to a certain area, intentional jamming in a given range, crosstalk avoidance in communication systems, selection of better signal-to-noise ratio and so on.

4.4. Advantages of using proposed CBRAM RF switch technology for reconfigurable antenna and filter applications

The proposed reconfigurable antenna and filter technology could be used for both applications which require an occasional reconfigurability or for systems which require a dynamic reconfigurability. Systems that require occasional reconfigurability include systems like Wi-Fi radiators, which aim to concentrate on a certain area of a building, or short distance communication antennas, which would be reconfigured only to cover a required range, like a centralized transponder for IoT applications addressing a group of smart devices. Occasionally reconfigurable systems also include passive microwave device applications, such as for energy harvesting, sensing, identification and so on. An example of a system that requires dynamic reconfigurability is an unmanned aerial vehicle (UAV) that needs to reorient its antenna dynamically to communicate with different base stations on earth, or a 5G communication link that adapts its antennas to achieve spatial and frequency diversity, or a communication link that selects a path with the best signal to loss ratio.

A good filter stage at the front end of a communication system would drastically cut down the complexity and budget of the required signal processing. The ability to reconfigure the filter and antenna part would help to adapt the same system ingeniously for a wide range of noises that occur occasionally due to various interruptions such as rain, cyclones, volcanic activity, power link crosstalk, electromagnetic noise from a construction site and so on. One could, in the long run, think of an efficient system capable of receiving a frequency hopping spread spectrum signal with an electronically frequency-reconfigurable filter and antenna system, synchronized with the transmitter using the transmission security key, to reconfigure the antenna and filter appropriately.

The emphasized innovation of the proposed devices in this chapter is their low- or zero-power operation (offered by non-volatile CBRAM/MIM switches), once a desired state is achieved by low-power pulses. Such dynamic and low-power electronically reconfigurable RF devices would be of great value in IoT and 5G applications. 5G data solutions entail more than just data speed, but allow secure and versatile control of remote machinery, sophisticated remote medical surgeries, driverless smart vehicles, and a multitude of smart passive or low-power stand-alone devices underlying the idea of the IoT (Borkar and Pande 2016; Collela 2017). All these above-mentioned applications require a “smart” switching system, with low or no power consumption, low cost, dynamic electronic reconfigurability and so on. The proposed CBRAM/MIM switches and the realized application examples such as reconfigurable filters, reconfigurable antennas, rewritable chipless RFID tags and so on usher in a new era of “passive connectivity” at an efficient budget.

Non-volatile MIM switches could enhance IoT and 5G applications both in terms of low-power stand-alone devices such as smart sensors, object identifiers and similar, and also the higher level data exchange devices such as transponders and communication links. 5G technologies open up requirements for a variety of beamforming and multichannel communication strategies. In all these systems, the relative consumption of power determines critically the final cost of distribution to the general public. Non-volatile MIM switches could help in reducing the power supply costs and also enhance system efficiency. 5G communication link requirements such as spatial diversity, tractable 3D beamforming, spatial multiplexing and interference cancellation applications could be implemented at a low establishment and operating cost, combining the features of CBRAM/MIM switches into advanced antenna and filter techniques, such as MIMO systems.

The possibility of realizing the proposed antenna system with switches as explained in Chapter 2, on flexible substrates, would also open a wide range of beamforming applications, which are generally limited by complicated shapes for a classic rigid substrate method of realization. With confidence from the presented example, the reconfigurable antenna based on a parasitic-loaded dipole, we can begin to think more about dynamically reconfigurable antennas. These could search for a communication signal or transmit selectively in a given direction, in all 360° of a given plane. This could be realized by printing the reconfigurable PEs on a flexible laminate, which then would be folded around to enclose a dipole antenna in the center or at a similar suitable position.

Such an antenna system could find applications across wide domains of radio frequency electronics. To give one simple application example, this antenna system could be used as a central controller transponder for a group of smart passive devices in IoT applications. In such a scenario, just by controlling the directional properties of an antenna radiation pattern, we could bring great relief to signal interference problems apparent in multitransponder environments. An illustration of this concept is shown in Figure 4.43. Here, a central feeding radiator is surrounded by a group of reconfigurable PEs with integrated MIM switches, fabricated on a flexible substrate, to fold around the radiator. The radiation pattern of this combination could be steered similarly to a reconfigurable transmit array by selectively switching the MIM switches for reconfiguring the PEs. Similar reconfigurable transmit array and reflect array concepts of a low cost, low weight and non-volatile nature could be sought with the proposed technology.

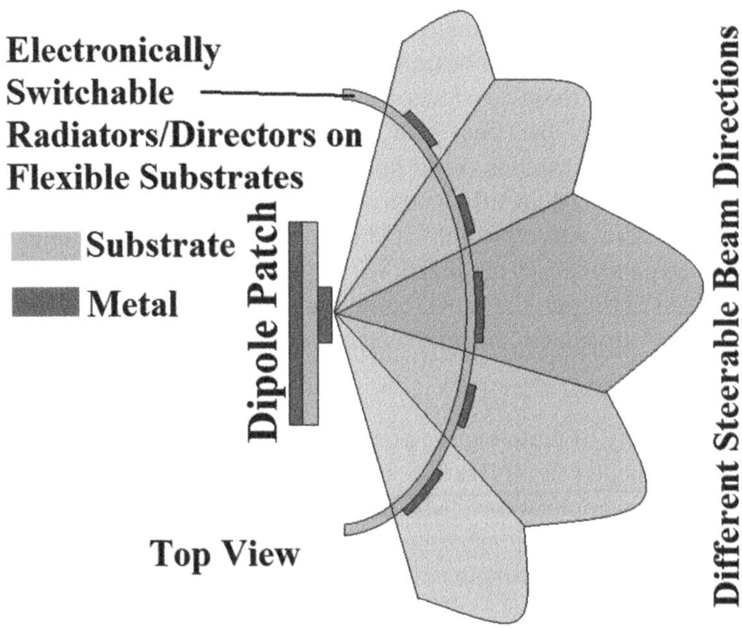

Figure 4.43. *Concept of flexible and electronically pattern steerable transmit array antenna system using non-volatile CBRAM/MIM RF switches. For a color version of this figure, see www.iste.co.uk/jayakrishnan/technology.zip*

It is also conceivable to use the proposed MIM switch technology for low-weight and flexible wearable RF circuit design applications. Rogier *et al.* (2018) report the integration of signal processing and power supply circuits on top of a metallic patch antenna, for compactness in wearable applications, without any degradation of antenna radiation performance. Similar designs could be realized using the proposed techniques for CBRAM/MIM RF switches and associated devices presented in this book by fabricating reconfigurable antenna and filter circuits on flexible substrates with adhesive background like sticker papers. These filters or other passive reconfigurable devices could be attached along with power supply circuits like solar cells on metallic parts of an antenna or on any suitable surface, layer by layer, to obtain a compact and flexible layout for a given application.

Last but not least, this very same switching technology could also be used to achieve frequency or polarization reconfigurability with an antenna. In such a case, the switches could be integrated in such a way that they tune the electrical length of the active radiating element to obtain a frequency reconfiguration. For polarization reconfigurability, the switches should be added to antenna geometry in a manner which reorients the current paths on the antenna surface so as to configure a desired polarization. These nonvolatile switches could also be designed to integrate into a switchable phase shifter network, which could feed a reconfigurable antenna array for pattern steering. Such a realization on flexible and low-cost substrates could bring significant benefits both in terms of operating power and cost of implementation.

4.5. Conclusion

In this chapter, we demonstrated the design and experimental realization of electronically reconfigurable filters and pattern steerable antennas. An insight into the operation of the presented filter topologies is given with the help of simple and efficient electrical model analysis. We extracted a mathematical expression that could be used for a quick design of the presented topology of band-stop filters. In addition, we also discussed the possible extension and enhancement techniques for the proposed reconfigurable filters for better and versatile performances.

Next, we presented an electronically reconfigurable antenna using an integrated MIM switch, showing a good non-volatile pattern steering performance, with acceptable gain and bandwidth characteristics. We showed the effect of MIM switch resistance on the antenna gain and gave a brief idea of the application areas of such non-volatile low-cost passive reconfigurable antennas.

A discussion of the advantages and possible future extension of the presented devices is also included in this chapter. Like for other devices presented in this study, these devices were also tested for 50 cycles of operation and were confirmed to be functional even after this limit test.

In general, using these proposed non ideal switches, one could easily add or short circuit a desired capacitance value to an RF path for reset (high impedance) and set (low impedance) states of the switch, respectively. A capacitance added in series to an RF path is directly linked to its electrical length. Thus, controlling the electrical length of an RF path in a given device to achieve electronic reconfigurability requires only a very simple yet effective technique.

Conclusion

We have presented the design and realization of a new generation of passive radio frequency (RF) and microwave devices based on non-volatile solid-state conductive bridging random access memory/metal insulator metal (CBRAM/MIM) RF switching technology. The presented RF switches and electronically reconfigurable RF and microwave devices using these integrated switches could benefit every field of RF electronics and especially as an important solution for low-cost stand-alone devices for the Internet of things (IoT) and fifth-generation cellular communication (5G) related applications.

The proposed switching technology and passive RF devices are fabricated using a low-cost and simple in-house process, which is compatible with mass productions in an industrial setup on both rigid and flexible substrates. The proposed techniques do not require a full-time "clean room" environment for their fabrication process; in the experiments related to this book a clean room support is utilized only to troubleshoot problems and benchmark the quality of the chosen realization process. Similarly, in an industrial environment, once a process is optimized for a device, it could be fabricated without any clean room support.

In the first chapter, we presented in detail the most relevant RF switching technologies at present, their principles of operation and modes of application, with the help of classified illustrations. The differences between classically used volatile switching solutions such as micro-electromechanical systems (MEMS) and solid-state semiconductor switches, and the non-volatile memristor-based switching solutions, phase change memory (PCM) and CBRAM/MIM, were clearly outlined. We also gave a detailed explanation, based on facts and references, of the requirements of a non-volatile switching technology, and its use in IoT and 5G applications. Finally, we clearly emphasized the reasons for the choice of CBRAM/MIM switches for this experiment, as opposed to to PCM switches, based on their simplicity in terms of application and operation.

In the second chapter, we presented the real-world implementation challenges of a low-cost non-volatile RF switching technology based on CBRAM/MIM switches. We also presented our efforts and outcomes in efficiently alleviating these challenges. Techniques for fabrication of CBRAM/MIM RF switches integrated into passive RF circuits, both on rigid as well as flexible substrates, at an economically efficient budget were discussed in this chapter. Design and realization of RF switches on 50 Ω CPW transmission lines, on classic FR-4 and flexible paper substrates, with good experimental RF performances were presented based on the optimized fabrication processes. Analysis and discussions of electrical equivalent models, the effect of filament resistance of CBRAM/MIM switches on RF performance, time stability and other interesting factors of the switch were detailed in the chapter. Finally, an application example of CPW-based RF switches for a single pole double throw power splitter was also presented affirmatively in this chapter.

In the third chapter, the applications of CBRAM/MIM non-volatile switches for electronically rewritable chipless RFID tags were presented with confirmed results. An introduction to chipless RFID technology, possible fields of application, reasons for referring to this technology as the "RF barcode of the future" and the reader technology used were explained. We presented the design and experimental realization of electronically rewritable chipless tags integrated with CBRAM/MIM switches, on rigid FR-4 and flexible paper substrates, with considerable performance. These realizations were based on the optimized fabrication process stated in Chapter 2. An insight into the working mechanism of these devices was explained with the help of an efficient electrical equivalent model. Discussions on data encoding strategies and potential advantages of using these electronically rewritable chipless RFID tags were also included in this chapter.

In the fourth chapter, fully passive solid-state electronically reconfigurable filter and antenna models based on CBRAM/MIM RF switches were presented. This realization was an investigation regarding the potential applications of the proposed CBRAM/MIM switch technology for classic RF device requirements, and was a proof of concept for the realization of non-volatile and electronically reconfigurable band stop filters and pattern steerable antennas. In this chapter, we explained the working mechanism of the reconfigurable filter with the help of an adequate electrical equivalent model. We also presented the techniques and some examples using full-wave electromagnetic simulations to extend the proposed RF switch technology to other filter topologies and overcome the limitations of presented designs. The presented pattern steerable antenna also showed impressive performance and was illustrated with the help of 2D and 3D experimentally obtained radiation patterns. A section of this chapter was dedicated to discussions related to the advantages of using the proposed CBRAM/MIM RF switches for reconfigurable antenna and filter applications, and the relevance of these in current technological advances in the field of IoT and 5G technologies.

Future studies related to this work and the findings presented in this book open up a new set of tasks aimed at optimizing the presented technology for real commercial applications. These include detailed studies on material aspects and RF design to improve the reliability and operating frequency upper limit of presented switches.

Studies related to more efficient and rapid fabrication of metallic and ion conductor layers for CBRAM/MIM switches through techniques like inkjet and thermal printing are also in progress, expected to deliver fully printed electronically reconfigurable and non-volatile passive RF devices. These studies will contribute to reducing the cost of realized devices and offer versatility such that a customer could print a desired complete electronically reconfigurable RF device using a common domestic printer.

We are also working on detailed and continued analysis in optimizing the presented devices and understanding their performance factors. This includes studies to determine features such as switching speed, optimization of the number of achievable switching cycles, analysis for more efficient utilization and handling of OFF state capacitances, dependence of device performance on environmental factors like humidity and temperature, and so on.

All these activities are carried out under a project called “ScattererID” funded by the Horizon 2020 (H2020) program of the European Research Council (ERC) of the European Union (EU) (grant agreement No. 772539) in a joint collaboration of the Laboratoire de Conception et d'Intégration des Systèmes (LCIS) of Grenoble Alpes University, and the Institut d'Electronique et des Systèmes (IES) of the University of Montpellier/ CNRS. Our efforts are also supported by the Centre National de la Recherche Scientifique (CNRS), France, and the Institut Universitaire de France (IUF), France.

Appendix

A.1. Observation of conductive filament formation in CBRAM/MIM switching cells

In this section, we present a control experiment that shows the conductive filament formation mechanism of copper-nafion-aluminum conductive bridging random access memory/metal insulator metal (CBRAM/MIM) devices. This is carried out using a generalized experiment setup, the details of which are explained in the following.

In the presented radio frequency (RF) applications in this book, the thickness of the ion conductor layer (nafion) in CBRAM/MIM switches (which is also the gap between metal electrodes) is chosen as 600 nm to assure an acceptable low impedance level (typically $< 25–30\ \Omega$) in set state as explained in Chapter 2. This thickness is optimized based on a previous study conducted among our group (Susanti 2016). However, such small thicknesses impose some limitations in the observation of a conductive filament formation in these devices. In addition, the layers formed in the above MIM cells are conformal to each other and block the view in classic optical transmission microscopes. Scanning electron microscope (SEM) observation of the mechanism in planar designs is also limited by the fact that the used polymer ion-conductors (nafion) leads to large surface charge accumulation on their layers from the electron beam and results in images without proper clarity.

To facilitate a better microscopic view of filament formation, we have modeled a control experiment utilizing the ion conductor properties of nafion itself for observation of the conductive filament dynamics in a macro (100 μm) switch gap. For this, we made a CBRAM/MIM switching cell on a glass plate with a 100 μm wide switch gap, as shown in Figure A.1. Switching action was observed in this MIM cell using an optical transmission microscope with a 10x magnification objective. The optically transparent nature of nafion makes such an observation feasible.

The experimental MIM switch is made by depositing nafion resin solution (ion conductor) on a standard microscopic glass slide of thickness 1 mm, carrying a layer of thermal vapor deposited aluminum (of thickness around 200 nm to 1 μm), which acts as the inert electrode. Nafion is placed as a drop on the edge of aluminum layer using a micropipette and measures around 200 μL. After the deposit of nafion, it is left to dry partially in an ambient environment to reach the consistency of a gel. Then we probe the nafion gel near the edge of aluminum layer to form an approximately 100 μm wide switch gap, using a copper wire which acts as the active electrode. The choice of a switch gap of 100 μm is based on the accuracy with which we could probe near to the aluminum edge using an available magnifier. In summary, we obtain an MIM cell in which an aluminum layer on glass is the inert electrode, nafion in gel form is the ion-conductor and a copper probe is the active electrode, as shown in Figure A.1. The probing of nafion and application of current pulses is done on the microscope platform, with the help of a specially designed two-axis movable secondary platform attachment.

The nafion resin solution used for the experiment is supplied by M/S Sigma Aldrich and contains nafion perflourinated resin solution – 5 wt.% in lower aliphatic alcohols and around 15–20% water (Sigma-Aldrich, 2019). It is a well-known fact that the ion conductivity of nafion is directly proportional to the water content of the molecule (Heitner-Wirguin 1996). The consistency of nafion is fixed by a trial and error method and is not technically quantified in this experiment.

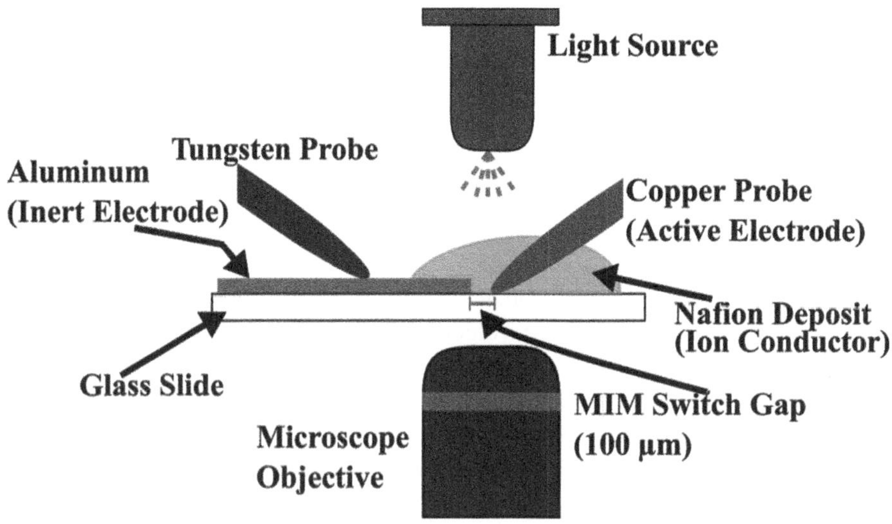

Figure A.1. *Illustration of experimental setup used for observation of conductive filament growth in copper-nafion-aluminum CBRAM/MIM switching cells using an optical transmission microscope. For a color version of this figure, see www.iste.co.uk/jayakrishnan/technology.zip*

However, a dry layer of nafion (like what is used for RF applications described in this book with a switching layer thickness of 600 nm) is not seen to show any ion conductivity suitable for observing the filament formation in a large gap (100 µm) as used here. Conversely, a layer of nafion with too much water content also hinders the view due to formation of a lot of bubbles, arising from electrolysis of water, upon application of the DC pulse used for operating the MIM switch. To observe the switching action and filament formation, we drive a DC current pulse through this MIM switch, using a Keithley 2400 source meter, and simultaneously record a microvideograph of the switch area using a high frame rate camera attached to the microscope. A photograph of this experimental setup is shown in Figure A.2.

A classic optical reflection-type microscope, which could give more clear color images, was not chosen for this experiment due to limitations in focusing on the switch area through the top side of glass slide shown in Figure A.1. The objective sizes we have (100×, 50×, 20×) with the optical reflection microscope cannot approach and focus on the switch area due to the presence of probes, probe holders and the large thickness of nafion drop.

We were also not able to use a magnification objective greater than 10× for the optical transmission microscope used, as it hits the glass slide in attempts to focus on the switching area. All these above-mentioned challenges could be understood simply by looking at the experimental setup shown in Figures A.1 and A.2. This issue could be partially overcome by using thinner glass slides, which will be included in future studies discussed at the end of this section.

The operating DC current pulses are applied with positive polarity with respect to the copper (active) electrode through the MIM switch. The magnitude of current was 0.5 mA and the compliance voltage limit was 10 V. Soon after the application of the current pulse, we could observe a dendrite-shaped filament growth from the inert electrode surface (aluminum layer) to the active electrode (copper probe), through the 100 μm wide ion conductor (nafion) layer.

The impedance of the MIM cell under test was initially (before application of set pulse) around 1–2 GΩ and gradually dropped to 10–20 kΩ on the complete formation of filament. The filament was fully formed in around 500 s. The photographs of different phases of filament formation are shown in Figure A.3.

Next, we tried to reset the cell using a current pulse of opposite polarity to the set pulse. The magnitude and compliance voltage value of the reset pulse was 200 mA and 3 V, respectively, and the resistance value was found to increase from 10 kΩ to 20 MΩ in the reset process.

However, no significant physical phenomenon was observed on the filament during reset process. We were able to set and reset the MIM switch for several cycles in the resistance range around 10 kΩ and 20 MΩ, respectively, for set and reset states. Also, in these experiments, no significant physical phenomenon was observed similar to the initial set (forming) process.

We managed to successfully repeat this process on several samples and the observations were similar. The above claims, that we did not observe any physical process of filament dissolution in the reset operation, or filament growth in successive set operations after the initial forming process, are justified based on the following hypothesis.

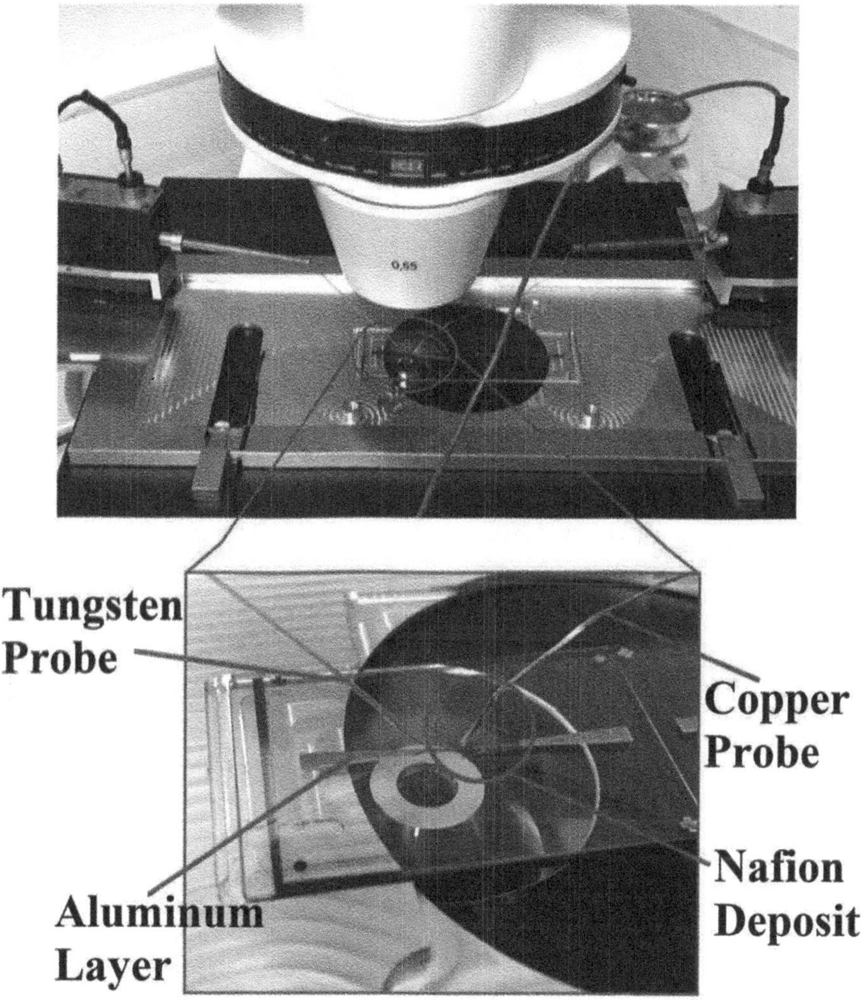

Figure A.2. *Photograph of experimental setup used for observation of conductive filament growth in copper-nafion-aluminum CBRAM/MIM switching cells using an optical transmission microscope. Inset shows zoomed photograph of MIM switch on glass slide. For a color version of this figure, see www.iste.co.uk/jayakrishnan/technology.zip*

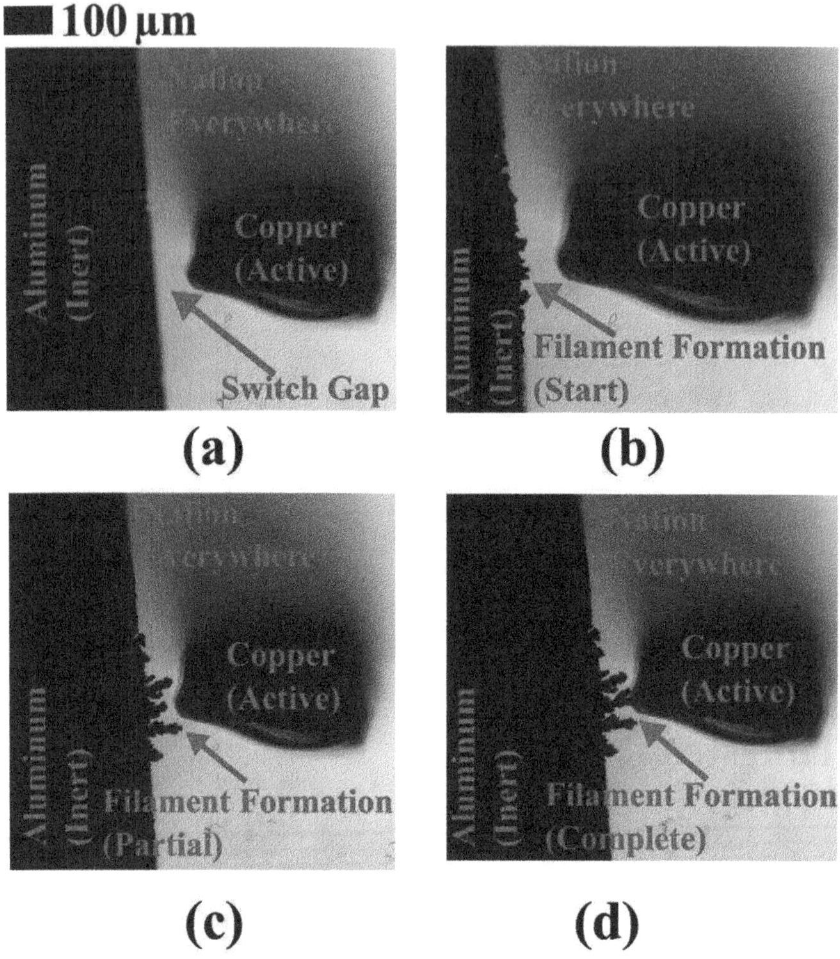

Figure A.3. *Microphotographs of different phase of filament formation in copper-nafion-aluminum CBRAM/MIM switching cells captured using an optical transmission microscope. For a color version of this figure, see www.iste.co.uk/jayakrishnan/technology.zip*

Firstly, the distribution of copper particles seen in close examination of the MIM switch cell after the experiment (after removing copper and tungsten probe), on a classic optical reflection microscope, showed a very granular appearance of the filament metal surface. Rather than a continuous metallic filament, a granular structure of distributed metallic particles was seen. These microphotographs are shown in Figures A.4 and A.5. This granular arrangement of copper particles, rather than a smooth continuous filament, is expected to be the reason for the observation of a very high set state resistance of the order of 10 kΩ.

Figure A.4. *Microphotographs of filament observed in forming process in copper-nafion-aluminum CBRAM/MIM switching cells, captured after removing the copper probe, using a classic optical microscope. For a color version of this figure, see www.iste.co.uk/jayakrishnan/technology.zip*

In addition, during the reset process, a rupture of the filament only in some weak parts of the metallic deposition, without a significant dissolution, is expected to take place. This rupture in filament leads the CBRAM/MIM cell to exhibit a high resistance state due to dissection of the current path. This rupture of conductive filament is expected to be of a very small scale, of the order of a few nanometers, and hence is not clearly resolved for observation through the present experimental setup.

Secondly, in the microscopic images of Figures A.4 and A.5 we see a staining in the nafion layer. This is expected to be due to heat produced in redox reactions, from prolonged application of the current pulse (>500 s).

This staining also hinders a free view of any mechanical deformation to the filament during the reset process.

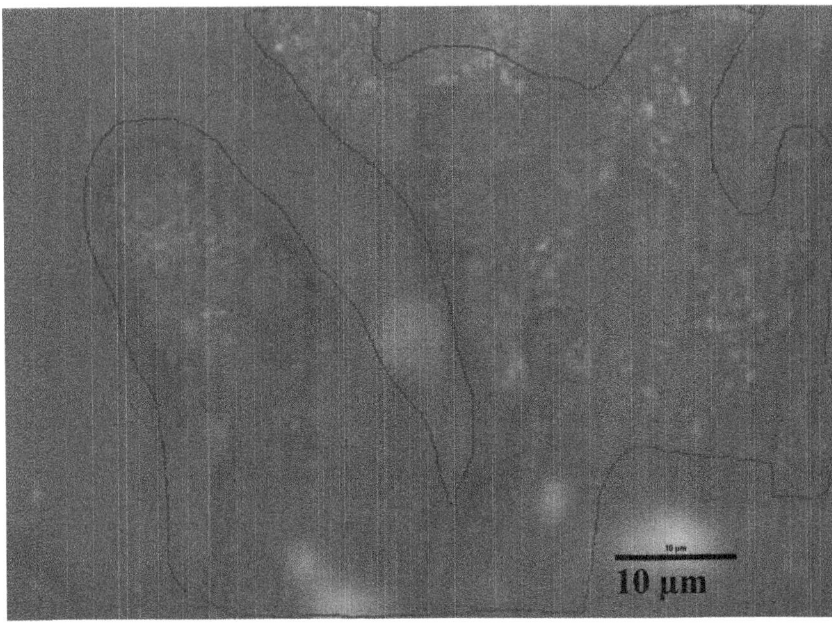

Figure A.5. *Zoomed microphotograph of filament observed in forming process in copper-nafion-aluminum CBRAM/MIM switching cells, showing the granular nature of metallic filaments, captured after removing the copper probe using a classic optical microscope. Filament patches are encircled in red. For a color version of this figure, see www.iste.co.uk/jayakrishnan/technology.zip*

Nevertheless, the primary aim of this experiment, to observe and validate the formation of a conductive filament in copper-nafion-aluminum CBRAM/MIM cells, was fulfilled successfully. Next, we hope to extract from these data the advanced characteristics of the proposed MIM switch, such as quantitative information on the ion conductivity of nafion, switch speed (speed of filament formation), resistance per unit length of filament for a given diameter and so on. Detailed analysis and more accurate experiments with reduced uncertainties in these aspects are in rapid progress among our group.

One generalized assumption is that higher set process currents would have led to stronger filaments and lower resistance states. Similarly higher reset state currents would have led to stronger dissolution of metallic filament. These values are not attempted at this stage due to the following reasons. If higher set state currents are used, it would obviously lead to faster and stronger filament formations; however, at the time, when a growing filament touches the opposite electrode, the overall resistance of the MIM switch is suddenly reduced. This would lead to higher current flow (which was initially limited due to higher resistance of the MIM cell) and thermal destruction of the formed filament due to its very thin dimensions.

Similarly, a reset pulse of higher magnitude would lead to localized dielectric breakdown or burning of the MIM cell. Both these events would hinder the intended purpose of this controlled experiment at this stage. These issues were observed in the initial trials of this experiment.

With the confidence from the obtained results we are planning a better characterization with a reduced number of uncertainties, such as benchmarking the water content in the nafion layer used, using a more efficient program to adapt the operating pulse for a safe regulation once a filament is formed to avoid its thermal destruction and so on.

In summary, the experiments presented here are just crude validations of the working mechanism of a copper-nafion-aluminum CBRAM/MIM switching cell, and the data and material properties were not quantified. Further analysis is required to benchmark these observations and draw a valid hypothesis and it is in rapid progress with the authors. Nevertheless, the observed physical phenomena validate the conductive bridging mechanism in copper–nafion–aluminum CBRAM/MIM switching cells. The direction of growth of metallic filament and the observed dynamics are in good agreement with scientifically valid articles like Yang *et al.* (2012). Experiments for more detailed and accurate observations in this context are planned as future work.

References

Adesto Technologies (2010). Adesto Technologies wins DARPA award to develop sub-threshold non-volatile, embedded CBRAM memory. Available: http://www.adestotech.com/news-detail/adesto-technologies-wins-darpa-award-to-develop-sub-threshold-non-volatile-embedded-cbram-memory/ [Accessed 4 July 2019].

Adesto Technologies (2016). RM24C32DS 32-Kbit 1.65V minimum non-volatile serial EEPROM I2C Bus. Available: https://www.adestotech.com/about-us/cbram/ [Accessed 9 January 2019].

Adesto Technologies (2020). CBRAM technology. Available: https://www.dialog-semiconductor.com/products/memory/cbram-technology [Accessed 4 July 2019].

Ali, Z., Siragusa, R., Perret, E., Barbot, N., Hely, D., Bernier, M., Garet, F. (2019). Méthode d'authentification basée sur des tags RFID sans puce imprimés par jet d'encre conductrice. *21èmes Journées Nationales Micro-Ondes (JNM2019)*. Caen, France.

Alves, A.A.C., Spadoti, D.H., Bravo-Roger, L.L. (2018). Optically controlled multiresonator for passive chipless tag. *IEEE Microwave and Wireless Components Letters*, 28(6), 467–469.

Al-Yasir, Y.I.A., Ojaroudi Parchin, N., Abd-Alhameed, R.A., Abdulkhaleq, A.M., Noras, J.M. (2019). Recent progress in the design of 4G/5G reconfigurable filters. *Electronics*, 8(1), 114.

Amin, E.M., Bhuiyan, M.S., Karmakar, N.C., Winther-Jensen, B. (2014a). Development of a low cost printable chipless RFID humidity sensor. *IEEE Sensors Journal*, 14(1), 140–149.

Amin, E.M., Saha, J.K., Karmakar, N.C. (2014b). Smart sensing materials for low-cost chipless RFID sensor. *IEEE Sensors Journal*, 14(7), 2198–2207.

Amin, M., Ramzan, R., Siddiqui, O. (2018). Slow wave applications of electromagnetically induced transparency in microstrip resonator. *Scientific Reports*, 8, 1–13.

Analog Devices (2018). 0 Hz/DC to 14 GHz, single-pole, four-throw MEMS switch with integrated driver. Available: https://www.analog.com/en/products/adgm1304.html [Accessed 25 June 2019].

Attaran, A., Rashidzadeh, R. (2016). Chipless radio frequency identification tag for IoT applications. *IEEE Internet of Things Journal*, 3(6), 1310–1318.

Bahl, I. (2003). *Lumped Elements for RF and Microwave Circuits*. Artech House, Inc. Norwood.

Balanis, C.A. (2005). *Antenna Theory: Analysis and Design*. John Wiley, Hoboken.

Bibi, F., Guillaume, C., Gontard, N., Sorli, B. (2017). A review: RFID technology having sensing aptitudes for food industry and their contribution to tracking and monitoring of food products. *Trends in Food Science & Technology*, 62, 91–103.

Bibi, F., Guillaume, C., Vena, A., Gontard, N., Sorli, B. (2016). Wheat gluten, a bio-polymer layer to monitor relative humidity in food packaging: Electric and dielectric characterization. *Sensors and Actuators A: Physical*, 247, 355–367.

Bocquet, M., Deleruyelle, D., Aziza, H., Muller, C., Portal, J., Cabout, T., Jalaguier, E. (2014). Robust compact model for bipolar oxide-based resistive switching memories. *IEEE Transactions on Electron Devices*, 61(3), 674–681.

Borgese, M., Dicandia, F.A., Costa, F., Genovesi, S., Manara, G. (2017). An inkjet printed chipless RFID sensor for wireless humidity monitoring. *IEEE Sensors Journal*, 17(15), 4699–4707.

Borkar, S., Pande, H. (2016). Application of 5G next generation network to Internet of Things. *2016 International Conference on Internet of Things and Applications (IOTA)*. IEEE, Pune, India, 443–447.

Borodulin, P., El-Hinnawy, N., Padilla, C.R., Ezis, A., King, M.R., Johnson, D.R., Nichols, D.T., Young, R.M. (2017). Recent advances in fabrication and characterization of GeTe-based phase-change RF switches and MMICs. *2017 IEEE MTT-S International Microwave Symposium (IMS)*. IEEE, Honolulu, USA, 285–288.

Bray, M.G., Werner, D.H. (2010). Passive switching of electromagnetic devices with memristors. *Applied Physics Letters*, 96(7), 073504.

Breed, G. (2010). A review of RF/microwave switching technologies. *High Frequency Electronics*, 9(5), 70–72.

Broadband Wireless Networking Lab (2019). 5G cellular systems: Broadband Wireless Networking Lab. Available: https://bwn.ece.gatech.edu/5G_systems/index.html [Accessed 8 July 2019].

Busby, Y., Nau, S., Sax, S., List-Kratochvil, E.J.W., Novak, J., Banerjee, R., Schreiber, F., Pireaux, J.-J. (2015). Direct observation of conductive filament formation in Alq3 based organic resistive memories. *Journal of Applied Physics*, 118(7), 075501.

Chan, K.Y., Daneshmand, M., Mansour, R.R., Ramer, R. (2007). Novel beam design for compact RF MEMS series switches. *2007 Asia-Pacific Microwave Conference*. IEEE, Bangkok, Thailand, 1–4.

Chan, K.Y., Ramer, R., Sorrentino, R. (2018). Low-cost Ku-band waveguide devices using 3-D printing and liquid metal filling. *IEEE Transactions on Microwave Theory and Techniques*, 66(9), 3993–4001.

Chan, R., Lesnick, R., Becher, D., Feng, M. (2003). Low-actuation voltage RF MEMS shunt switch with cold switching lifetime of seven billion cycles. *Journal of Microelectromechanical Systems*, 12(5), 713–719.

Chanthbouala, A., Garcia, V., Cherifi, R.O., Bouzehouane, K., Fusil, S., Moya, X., Xavier, S., Yamada, H., Deranlot, C., Mathur, N.D., Bibes, M., Barthélémy, A., Grollier, J. (2012). A ferroelectric memristor. *Nature Materials*, 11(10), 860–864.

Chaudhary, G., Jeong, Y., Lim, J. (2012). Harmonic suppressed dual-band bandpass filters with tunable passbands. *IEEE Transactions on Microwave Theory and Techniques*, 60(7), 2115–2123.

Chaudhary, G., Jeong, Y., Lim, J. (2013). Dual-band bandpass filter with independently tunable center frequencies and bandwidths. *IEEE Transactions on Microwave Theory and Techniques*, 61(1), 107–116.

Chen, S.L., Qin, P.Y., Lin, W., Guo, Y.J. (2018). Pattern reconfigurable antenna with five switchable beams in elevation plane. *IEEE Antennas and Wireless Propagation Letters*, PP(99), 1–1.

Chen, Z., Zhang, S., Chu, Q. (2017). Dual-band reconfigurable bandstop filter with independently controlled stopbands and constant absolute bandwidths. *2017 IEEE MTT-S International Microwave Symposium (IMS)*, Honololu, USA, 926–928.

Chua, L. (1971). Memristor – The missing circuit element. *IEEE Transactions on Circuit Theory*, 18(5), 507–519.

Chua, L. (2011). Resistance switching memories are memristors. *Applied Physics A*, 102(4), 765–783.

Collela, P. (2017). 5G and IoT: Ushering in a new era. Available: https://www.ericsson.com/en/about-us/india/authored-articles/5g-and-iot-ushering-in-a-new-era [Accessed 8 July 2019].

Cook, B.S., Le, T., Palacios, S., Traille, A., Tentzeris, M.M. (2013). Only skin deep: Inkjet-printed zero-power sensors for large-scale RFID-integrated smart skins. *IEEE Microwave Magazine*, 14(3), 103–114.

Daughton, J.M., Huang, J.S.T. (1988). Magnetoresistive memory including thin film storage cells having tapered ends. Patent, US4731757A. Available: https://patents.google.com/patent/US4731757A/en [Accessed 28 June 2019].

Ding, X., Wang, B.Z. (2013). A novel wideband antenna with reconfigurable broadside and endfire patterns. *IEEE Antennas and Wireless Propagation Letters*, 12, 995–998.

Dragoman, M., Aldrigo, M., Adam, G. (2017). Phased antenna arrays based on non-volatile resistive switches. *Antennas Propagation IET Microwaves*, 11(8), 1169–1173.

Dragoman, M., Cismaru, A., Hartnagel, H., Plana, R. (2006). Reversible metal-semiconductor transitions for microwave switching applications. *Applied Physics Letters*, 88(7), 073503.

El-Hinnawy, N., Borodulin, P., Wagner, B.P., King, M.R., Jones, E.B., Howell, R.S., Lee, M.J., Young, R.M. (2014). Low-loss latching microwave switch using thermally pulsed non-volatile chalcogenide phase change materials. *Applied Physics Letters*, 105(1), 013501.

Ellinger, F., Ishida, K., Meister, T., Boroujeni, B.K., Barahona, M., Carta, C., Münzenrieder, N., Cantarella, G., Petti, L., Knobelspies, S., Salvatore, G.A., Tröster, G., Schmidt, G.C., Hübler, A.C. (2018). Bendable printed and thin-film electronics for wireless communications. *2018 2nd URSI Atlantic Radio Science Meeting (AT-RASC)*. IEEE, Meloneras, Spain, 1–4.

Ellinger, F., Ishida, K., Shabanpour, R., Meister, T., Boroujeni, B.K., Carta, C., Petti, L., Salvatore, G.A., Tröster, G., Münzenrieder, N. (2015). Radio frequency electronics on plastic. *2015 SBMO/IEEE MTT-S International Microwave and Optoelectronics Conference (IMOC)*. IEEE, Porto de Galinhas, Brazil, 1–5.

Farzami, F., Khaledian, S., Smida, B., Erricolo, D. (2017). Pattern-reconfigurable printed dipole antenna using loaded parasitic elements. *IEEE Antennas and Wireless Propagation Letters*, 16, 1151–1154.

Feng, W., Shang, Y., Che, W., Gómez-García, R., Xue, Q. (2017). Multifunctional reconfigurable filter using transversal signal-interaction concepts. *IEEE Microwave and Wireless Components Letters*, 27(11), 980–982.

Feng, Y., Xie, L., Chen, Q., Zheng, L. (2015). Low-cost printed chipless RFID humidity sensor tag for intelligent packaging. *IEEE Sensors Journal*, 15(6), 3201–3208.

Finkenzeller, K. (2010). *RFID Handbook: Fundamentals and Applications in Contactless Smart Cards, Radio Frequency Identification and Near-Field Communication*. John Wiley and Sons, Hoboken.

Garbati, M., Perret, E., Siragusa, R., Halope, C. (2019). Ultrawideband chipless RFID: Reader technology from SFCW to IR-UWB. *IEEE Microwave Magazine*, 20(6), 74–88.

Gawelda, W., Siegel, J., Afonso, C.N., Plausinaitiene, V., Abrutis, A., Wiemer, C. (2011). Dynamics of laser-induced phase switching in GeTe films. *Journal of Applied Physics*, 109(12), 123102.

Goldsmith, C.L., Yao, Z., Eshelman, S., Denniston, D. (1998). Performance of low-loss RF MEMS capacitive switches. *IEEE Microwave and Guided Wave Letters*, 8(8), 269–271.

Gregory, M.D., Werner, D.H. (2015). Application of the memristor in reconfigurable electromagnetic devices. *IEEE Antennas and Propagation Magazine*, 57(1), 239–248.

Guy, J. (2015). Evaluation des performances des mémoires CBRAM (Conductive bridge memory) afin d'optimiser les empilements technologiques et les solutions d'intégration. PhD Thesis, Grenoble Alpes University. Available: http://www.theses.fr/2015GREAT127 [Accessed 3 July 2019].

Havens, R.D., Japp, R.M., Knight, J.A., Poliks, M.D., Quinn, A.M. (2002). Electronic package utilizing protective coating. Patent, US6351030B2. Available: https://patents.google.com/patent/US6351030B2/en [Accessed 21 March 2019].

Heitner-Wirguin, C. (1996). Recent advances in perfluorinated ionomer membranes: Structure, properties and applications. *Journal of Membrane Science*, 120(1), 1–33.

Hillman, C., Stupar, P.A., Hacker, J.B., Griffith, Z., Field, M., Rodwell, M. (2014). An ultra-low loss millimeter-wave solid state switch technology based on the metal-insulator-transition of vanadium dioxide. *2014 IEEE MTT-S International Microwave Symposium (IMS2014)*. IEEE, Tampa, USA, 1–4.

Hong, J. (2009). Reconfigurable planar filters. *IEEE Microwave Magazine*, 10(6), 73–83.

Islam, M.A., Karmakar, N.C. (2015). Real-world implementation challenges of a novel dual-polarized compact printable chipless RFID tag. *IEEE Transactions on Microwave Theory and Techniques*, 63(12), 4581–4591.

Islam, Md.A., Karmakar, N.C. (2012). A novel compact printable dual-polarized chipless RFID system. *IEEE Transactions on Microwave Theory and Techniques*, 60(7), 2142–2151.

ITRS Reports (2011). International Technology Roadmap for Semiconductors. Available: http://www.itrs2.net/itrs-reports.html [Accessed 4 July 2019].

Jaffe, M., Abou-Khalil, M., Botula, A., Ellis-Monaghan, J., Gambino, J., Gross, J., He, Z., Joseph, A., Phelps, R., Shank, S., Slinkman, J., Wolf, R. (2015). Improvements in SOI technology for RF switches. *2015 IEEE 15th Topical Meeting on Silicon Monolithic Integrated Circuits in RF Systems*. IEEE, San Diego, USA, 30–32.

Jalaly, I., Robertson, I.D. (2005). Capacitively-tuned split microstrip resonators for RFID barcodes. *2005 European Microwave Conference*. Paris, France.

Jayakrishnan, M.P., Vena, A., Meghit, A., Sorli, B., Perret, E. (2017). Nafion-based fully passive solid-state conductive bridging RF switch. *IEEE Microwave and Wireless Components Letters*, 27(12), 1104–1106.

Jayakrishnan, M.P., Vena, A., Sorli, B., Perret, E. (2018a). Solid-state conductive-bridging reconfigurable RF-encoding particle for chipless RFID applications. *IEEE Microwave and Wireless Components Letters*, 28(6), 506–508.

Jayakrishnan, M.P., Vena, A., Sorli, B., Perret, E. (2018b). Electronically re-configurable, non-volatile, nano-ionics-based rf-switch on paper substrate for chipless RFID applications. *Technologies*, 6(3), 58.

Jayakrishnan, M.P., Perret, E., Vena, A., Sorli, B. (2019a). Electronically reconfigurable dipole antenna using integrated passive non-volatile solid-state metal-insulator-metal switches. *IEEE International Symposium on Antennas and Propagation and USNC-URSI National Radio Science Meeting 2019*. IEEE, Atlanta, 1–2.

Jayakrishnan, M.P., Vena, A., Sorli, B., Perret, E. (2019b). Investigation of integrated solid state nano-ionic metal-insulator-metal switches for electronically reconfigurable band stop filter applications. *IET Microwaves, Antennas & Propagation*, 13(12), 1963–1968.

Jian, J., Chang, H., Vena, A., Sorli, B. (2017). Study and design of resistive switching behaviors in PMMA-based conducting-bridge random-access memory (CBRAM) devices. *Microsystem Technologies*, 23(6), 1719–1725.

Jordan, E.C., Balmain, K.G. (1968). *Electromagnetic Waves and Radiating Systems.* Prentice Hall, New Jersey.

Jung, C.W., Lee, M.J., Li, G.P., Flaviis, F.D. (2006). Reconfigurable scan-beam single-arm spiral antenna integrated with RF-MEMS switches. *IEEE Transactions on Antennas and Propagation*, 54(2), 455–463.

Karmaker, N.C. (2016). Tag, you're it radar cross section of chipless RFID tags. *IEEE Microwave Magazine*, 17(7), 64–74.

Keysight Technologies (2014). RF and microwave switching solutions. Selection guide.

Keysight Technologies (2017a). Keysight U9397A/C FET solid state switch (SPDT).

Keysight Technologies (2017b). Understanding RF/microwave solid state switches and their applications. Application note.

Keysight Technologies (2019). P9400C solid state PIN diode transfer switch, 100 MHz to 18 GHz. Available: https://www.keysight.com/en/pd-1205162-pn-P9400C/solid-state-pin-diode-transfer-switch-100-mhz-to-18-ghz?cc=FR&lc=fre [Accessed 25 June 2019].

Kozicki, M.N. (2002). Programmable microelectronic devices and method of forming and programming same. Patent, US6487106B1. Available: https://patents.google.com/patent/US6487106/en [Accessed 4 July 2019].

Kozicki, M.N. (2010). Programmable metallization cell structures including an oxide electrolyte, devices including the structure and method of forming same. Patent, US7728322B2. Available: https://patents.google.com/patent/US7728322/en [Accessed 4 July 2019].

Kozicki, M.N., Balakrishnan, M. (2008). Programmable metallization cell structures including an oxide electrolyte, devices including the structure and method of forming same. Patent, US7372065B2. Available: https://patents.google.com/patent/US7372065/en [Accessed 4 July 2019].

Kozicki, M.N., West, W.C. (1998). Programmable metallization cell structure and method of making same. Patent, U.S. Patent 5,761,115. Available: https://patents.google.com/patent/US5761115A/en [Accessed 4 May 2018].

Kozicki, M.N., West, W.C. (2002). Programmable sub-surface aggregating metallization structure and method of making same. Patent, US6418049B1. Available: https://patents.google.com/patent/US6418049/en [Accessed 4 July 2019].

Lee, J., Je, C.-H., Kang, S., Choi, C.-A. (2005). A single-pole 6-throw (SP6T) antenna switch using metal-contact RF MEMS switches for multi-band applications. *IEEE MTT-S International Microwave Symposium Digest*, IEEE, Long Beach, USA.

Lee, T., Lee, S. (2010). Modeling of SOI FET for RF switch applications. *2010 IEEE Radio Frequency Integrated Circuits Symposium*. IEEE, Anaheim, USA, 479–482.

Leon, A. (2018). Commutateurs RF à base de matériaux à changement d'état: conception, caractérisation et application. PhD Thesis, Grenoble Alpes University. Available: http://www.theses.fr/s147914 [Accessed 3 July 2019].

Li, P., Cui, T. (2015). RF nano switch based on single crystalline graphene. *18th International Conference on Solid-State Sensors, Actuators and Microsystems (TRANSDUCERS)*. IEEE, Anchorage, USA, 307–310.

Li, Q., Zhang, Y.P., Yeo, K.S., Lim, W.M. (2008). 16.6- and 28-GHz fully integrated CMOS RF switches with improved body floating. *IEEE Transactions on Microwave Theory and Techniques*, 56(2), 339–345.

Li, R., Wu, T., Pan, B., Lim, K., Laskar, J., Tentzeris, M.M. (2009). Equivalent-circuit analysis of a broadband printed dipole with adjusted integrated balun and an array for base station applications. *IEEE Transactions on Antennas and Propagation*, 57(7), 2180–2184.

Lim, S., Ling, H. (2007). Design of electrically small, pattern reconfigurable Yagi antenna. *Electronics Letters*, 43(24), 1326–1327.

Lin, K.-L., Hou, T.-H., Shieh, J., Lin, J.-H., Chou, C.-T., Lee, Y.-J. (2011). Electrode dependence of filament formation in HfO2 resistive-switching memory. *Journal of Applied Physics*, 109(8).

Lin, W., Wong, H., Ziolkowski, R.W. (2017). Wideband pattern-reconfigurable antenna with switchable broadside and conical beams. *IEEE Antennas and Wireless Propagation Letters*, 16, 2638–2641.

Liu, S., Ji, Y., Yang, J., Sun, W., Li, H. (2018). Nafion film temperature/humidity sensing based on optical fiber Fabry-Perot interference. *Sensors and Actuators A: Physical*, 269, 313–321.

Liu, H., Wang, Z.-H., Li, L., Fan, Y.-X., Tao, Z.-Y. (2019). Vanadium dioxide-assisted broadband tunable terahertz metamaterial absorber. *Scientific Reports*, 9(1), 5751.

Lu, Z., Lanagan, M., Manias, E., Macdonald, D.D. (2009). Two-port transmission line technique for dielectric property characterization of polymer electrolyte membranes. *The Journal of Physical Chemistry B*, 113(41), 13551–13559.

Lu, Z.L., Yang, X.X., Tan, G.N. (2017). A multidirectional pattern-reconfigurable patch antenna with CSRR on the ground. *IEEE Antennas and Wireless Propagation Letters*, 16, 416–419.

Ma, X., Hong, W., Li, S., Yin, L., Deng, L. (2016). A terahertz RF selective switch based on graphene. *2016 IEEE International Conference on Internet of Things (iThings) and IEEE Green Computing and Communications (GreenCom) and IEEE Cyber, Physical and Social Computing (CPSCom) and IEEE Smart Data (SmartData)*. IEEE, Chengdu, China, 282–286.

Mahanta, P., Munna, M., Coutu, R.A. (2018). Performance comparison of phase change materials and metal-insulator transition materials for direct current and radio frequency switching applications. *Technologies*, 6(2), 48.

Makimoto, M., Yamashita, S. (2001). *Microwave Resonators and Filters for Wireless Communication: Theory, Design and Application*. Springer Science and Business Media, Berlin.

Mizutani, H., Takayama, Y. (2000). DC-110-GHz MMIC traveling-wave switch. *IEEE Transactions on Microwave Theory and Techniques*, 48(5), 840–845.

Moldovan, C.F., Vitale, W.A., Tamagnone, M., Ionescu, A.M. (2015). Graphene RF NEMS shunt switches for analog and digital phase shifters. *18th International Conference on Solid-State Sensors, Actuators and Microsystems (TRANSDUCERS)*. IEEE, Anchorage, USA, 2029–2032.

Mollah, M.N., Karmakar, N.C. (2001). RF-MEMS switches: Paradigms of microwave switching. *Asia-Pacific Microwave Conference (APMC)*. IEEE, Taipei, Taiwan, 1024–1027.

Moon, J., Seo, H., Son, K., Lee, K., Zehnder, D., Tai, H., Le, D. (2018). Phase-change RF switches with robust switching cycle endurance. *2018 IEEE Radio and Wireless Symposium (RWS)*. IEEE, Anaheim, USA, 231–233.

MVG (2014). MVG StarLab antenna measurement system. Available: https://www.mvg-world.com/en/products/field_product_family/antenna-measurement-2/starlab [Accessed 17 June 2019].

Nair, R.S., Perret, E., Tedjini, S., Baron, T. (2013). A group-delay-based chipless RFID humidity tag sensor using silicon nanowires. *IEEE Antennas and Wireless Propagation Letters*, 12, 729–732.

Nessel, J., Lee, R. (2011). *Chalcogenide nanoionic-based radio frequency switch*. Patent, US 7.923,715 B2. Available: https://patents.google.com/patent/US7923715B2/en [Accessed 7 May 2018].

Nessel, J.A., Lee, R.Q., Mueller, C.H., Kozicki, M.N., Minghan Ren, Morse, J. (2008). A novel nanoionics-based switch for microwave applications. *2008 IEEE MTT-S International Microwave Symposium Digest*. IEEE, Atlanta, USA, USA, 1051–1054.

Nijas, C.M., Dinesh, R., Deepak, U., Rasheed, A., Mridula, S., Vasudevan, K., Mohanan, P. (2012). Chipless RFID tag using multiple microstrip open stub resonators. *IEEE Transactions on Antennas and Propagation*, 60(9), 4429–4432.

Oberhammer, J. (2017). THz MEMS—Micromachining enabling new solutions at millimeter and submillimeter-wave frequencies. *2017 IEEE Asia Pacific Microwave Conference (APMC)*. IEEE, Kuala Lumpar, Malaysia, 81–84.

Pagnia, H., Sotnik, N. (1988). Bistable switching in electroformed metal–insulator–metal devices. *Physica Status Solidi (a)*, 108(1), 11–65.

Pan, K., Leng, T., Zhang, X., Hu, Z. (2017). Design and modeling of back gated graphene based RF switch with CPW transmission line on a high resistivity silicon substrate. *2017 10th UK-Europe-China Workshop on Millimetre Waves and Terahertz Technologies (UCMMT)*. IEEE, Liverpool, UK, 1–2.

Perret, E. (2014). *Radio Frequency Identification and Sensors: From RFID to Chipless RFID*. ISTE Ltd, London and John Wiley and Sons, Hoboken.

Perret, E. (2017). Micrometrie displacement sensor based on chipless RFID. *2017 IEEE MTT-S International Microwave Symposium (IMS)*. IEEE, Honololu, USA, 605–608.

Perret, E., Tedjini, S., Vasudevan, N.D., Garet, F., Duvillaret, L., Vena, A.M. (2013). Chipless passive RFID tag. Patent, US8556184B2. Available: https://patents.google.com/patent/US8556184/en [Accessed 19 February 2019].

Perret, E., Vidal, T.L., Vena, A., Gonon, P. (2015). Realization of a conductive bridging RF switch integrated onto printed circuit board. *Progress in Electromagnetics Research*, 151, 9–16.

Pi, S., Ghadiri-Sadrabadi, M., Bardin, J.C., Xia, Q. (2015). Nanoscale memristive radiofrequency switches. *Nature Communications*, 6, 7519.

Pi, S., Ghadiri-Sadrabadi, M., Bardin, J.C., Xia, Q. (2016). Memristors as radiofrequency switches. *2016 IEEE International Symposium on Circuits and Systems (ISCAS)*. IEEE, Montreal, Canada, 377–380.

Plessky, V.P., Reindl, L.M. (2010). Review on SAW RFID tags. *IEEE Transactions on Ultrasonics, Ferroelectrics, and Frequency Control*, 57(3), 654–668.

Pozar, D.M. (2005). *Microwave Engineering*. John Wiley and Sons, Hoboken.

Preradovic, S., Balbin, I., Karmakar, N.C., Swiegers, G.F. (2009). Multiresonator-based chipless RFID system for low-cost item tracking. *IEEE Transactions on Microwave Theory and Techniques*, 57(5), 1411–1419.

Preradovic, S., Karmakar, N.C. (2010). Chipless RFID: bar code of the future. *IEEE Microwave Magazine*, 11(7), 87–97.

Psychogiou, D., Gómez-García, R., Peroulis, D. (2016). Fully-reconfigurable bandpass/bandstop filters and their coupling-matrix representation. *IEEE Microwave and Wireless Components Letters*, 26(1), 22–24.

Raieszadeh, M., Shim, Y., Wang, M. (2016). *Directly heated RF phase change switch*. Patent, WO2016028362A2. Available: https://patents.google.com/patent/WO2016028362A2/en [Accessed 28 June 2019].

Ramos, A., Perret, E., Rance, O., Tedjini, S., Lázaro, A., Girbau, D. (2016). Temporal separation detection for chipless depolarizing frequency-coded RFID. *IEEE Transactions on Microwave Theory and Techniques*, 64(7), 2326–2337.

Rebeiz, G.M. (2003). RF MEMS switches: Status of the technology. *12th International Conference on Solid-State Sensors, Actuators and Microsystems*. IEEE, Boston, USA. 1726–1729.

Rebeiz, G.M. (2004). *RF MEMS: Theory, Design, and Technology*. John Wiley and Sons, Hoboken.

Rebeiz, G.M., Guan-Leng Tan, Hayden, J.S. (2002). RF MEMS phase shifters: Design and applications. *IEEE Microwave Magazine*, 3(2), 72–81.

Rogier, H., Van Baelen, D., Lemey, S., Agneessens, S., Rossi, M., Verhaevert, J. (2018). Some techniques for mass production of textile wireless systems. *2nd URSI Atlantic Radio Science Meeting (AT-RASC)*. 1–4. Available: http://hdl.handle.net/1854/LU-8568103 [Accessed 17 June 2019].

Sadiku, M.N.O. (2007). *Elements of Electromagnetics*. Oxford University Press, Oxford.

Saggin, B., Belaizi, Y., Vena, A., Sorli, B., Guillard, V., Dedieu, I. (2019). A flexible biopolymer based UHF RFID-sensor for food quality monitoring. *2019 IEEE International Conference on RFID Technology and Applications (RFID-TA)*. IEEE, Pisa, Italy, 484–487.

Saracho-Pantoja, I.O., Montejo-Garai, J.R., Rebollar, J.A.R.-C. and J.M. (2018). Additive manufacturing of 3D printed microwave passive components. *Emerging Microwave Technologies in Industrial, Agricultural, Medical and Food Processing*. [Online]. Available: https://www.intechopen.com/books/emerging-microwave-technologies-in-industrial-agricultural-medical-and-food-processing/additive-manufacturing-of-3d-printed-microwave-passive-components [Accessed 9 July 2019].

Shim, Y., Hummel, G., Rais-Zadeh, M. (2013). RF switches using phase change materials. *2013 IEEE 26th International Conference on Micro Electro Mechanical Systems (MEMS)*. IEEE, Taipei, Taiwan, 237–240.

Sigma-Aldrich (2019). *Nafion® perfluorinated resin solution*. Available: https://www.sigmaaldrich.com/catalog/product/aldrich/274704 [Accessed 21 March 2019].

Song, M.-J., Kwon, K.-H., Park, J.-G. (2017). Electro-forming and electro-breaking of nanoscale Ag filaments for conductive-bridging random-access memory cell using Ag-doped polymer-electrolyte between Pt electrodes. *Scientific Reports*, 7(1), 3065.

Souchon, F., Saint-Patrice, D., Pornin, J.L., Bouchu, D., Baret, C., Reig, B. (2017). Thin film packaged redundancy RF MEMS switches for space applications. *19th International Conference on Solid-State Sensors, Actuators and Microsystems (TRANSDUCERS)*. IEEE, Kaohsiung, Taiwan, 175–178.

Storm, H.F. (1967). Field-effect transistor (FET) bibliography. *IEEE Transactions on Electron Devices*, 14(10), 710–717.

Strukov, D.B., Snider, G.S., Stewart, D.R., Williams, R.S. (2008). The missing memristor found. *Nature*, 453(7191), 80–83.

Susanti, I. (2016). Study and design of non-volatile conductive bridge memory devices on flexible substrate. Master's Thesis, Institut Teknologi Bandung. Available: http://eprints.itb.ac.id/21/ [Accessed 16 July 2019].

Suzuki, K. (2001). MEMS micro-switches for use in DC, RF, and optical applications. *International Microprocesses and Nanotechnology Conference*. IEEE, Shimane, Japan, 48–49.

Tedjini, S., Karmakar, N., Perret, E., Vena, A., Koswatta, R., E-Azim, R. (2013). Hold the chips: Chipless technology, an alternative technique for RFID. *IEEE Microwave Magazine*, 14(5), 56–65.

Unlu, M., Jarrahi, M. (2014). Miniature multi-contact MEMS switch for broadband terahertz modulation. *Optics Express*, 22(26), 32245–32260.

Vena, A., Perret, E., Tedjini, S. (2011). Chipless RFID tag using hybrid coding technique. *IEEE Transactions on Microwave Theory and Techniques*, 59(12), 3356–3364.

Vena, A., Perret, E., Tedjini, S., Vallée, C., Gonon, P., Mannequin, C. (2012a). A fully passive RF switch based on nanometric conductive bridge. *2012 IEEE/MTT-S International Microwave Symposium Digest*. IEEE, Montreal, Canada, 1–3.

Vena, A., Perret, E., Tedjini, S. (2012b). A fully printable chipless RFID tag with detuning correction technique. *IEEE Microwave and Wireless Components Letters*, 22(4), 209–211.

Vena, A., Babar, A.A., Sydanheimo, L., Tentzeris, M.M., Ukkonen, L. (2013a). A novel near-transparent ASK-reconfigurable inkjet-printed chipless RFID tag. *IEEE Antennas and Wireless Propagation Letters*, 12, 753–756.

Vena, A., Perret, E., Tedjini, S. (2013b). Design rules for chipless RFID tags based on multiple scatterers. *Annals of Telecommunications*, 68(7–8), 361–374.

Vena, A., Perret, E., Tedjini, S. (2016). *Chipless RFID Based on RF Encoding Particle: Realization, Coding and Reading System*. ISTE Press, London and Elsevier, New York.

Vena, A., Sorli, B., Belaizi, Y., Saggin, B., Podlecki, J. (2018). An inkjet printed RFID-enabled humidity sensor on paper based on biopolymer. *2018 2nd URSI Atlantic Radio Science Meeting (AT-RASC)*. IEEE, Meloneras, Spain, 1–3.

Wang, L., Yuan, M., Xiao, T., Joines, W.T., Liu, Q.H. (2011). Broadband electromagnetic radiation modulated by dual memristors. *IEEE Antennas and Wireless Propagation Letters*, 10, 623–626.

Wang, M., Shim, Y., Rais-Zadeh, M. (2014). A low-loss directly heated two-port RF phase change switch. *IEEE Electron Device Letters*, 35(4), 491–493.

Wu, R.-J., Sun, Y.-L., Lin, C.-C., Chen, H.-W., Chavali, M. (2006). Composite of TiO_2 nanowires and nafion as humidity sensor material. *Sensors and Actuators B: Chemical*, 115(1), 198–204.

Wuttig, M., Yamada, N. (2007). Phase-change materials for rewriteable data storage. *Nature Materials*, 6(11), 824–832.

Xia, Q., Bardin, J., Pi, S. (2015). *Memristive RF switches*. Patent, US20150303376A1. Available: https://patents.google.com/patent/US20150303376/en [Accessed 5 July 2019].

Yalon, E., Karpov, I., Karpov, V., Riess, I., Kalaev, D., Ritter, D. (2015). Detection of the insulating gap and conductive filament growth direction in resistive memories. *Nanoscale*, 7(37), 15434–15441.

Yang, Y., Gao, P., Gaba, S., Chang, T., Pan, X., Lu, W. (2012). Observation of conducting filament growth in nanoscale resistive memories. *Nature Communications*, 3, 732.

Yang, Y., Gao, P., Li, L., Pan, X., Tappertzhofen, S., Choi, S., Waser, R., Valov, I., Lu, W.D. (2014). Electrochemical dynamics of nanoscale metallic inclusions in dielectrics. *Nature Communications*, 5, 4232.

Zhang, S., Zhu, L. (2011). Compact and high-selectivity microstrip bandpass filters using triple-/quad-mode stub-loaded resonators. *IEEE Microwave and Wireless Components Letters*, 21(10), 522–524.

Zheng, L., Rodriguez, S., Zhang, L., Shao, B., Zheng, L.-R. (2008). Design and implementation of a fully reconfigurable chipless RFID tag using inkjet printing technology. *IEEE International Symposium on Circuits and Systems*. IEEE, Seattle, WA, 1524–1527.

Zohur, A., Mopidevi, H., Rodrigo, D., Unlu, M., Jofre, L., Cetiner, B.A. (2013). RF MEMS reconfigurable two-band antenna. *IEEE Antennas and Wireless Propagation Letters*, 12, 72–75.

Index

Other titles from

iSTE

in

Electronics Engineering

2021

NGUYEN Thien-Phap
Organic Electronics 1: Materials and Physical Processes
Organic Electronics 2: Applications and Marketing

2020

CAPPY Alain
Neuro-inspired Information Processing

FOUQUET François
Noise in Radio-Frequency Electronics and its Measurement

MAHDOUM Ali
CAD of Circuits and Integrated Systems

2018

MURET Pierre
Fundamentals of Electronics 2: Continuous-time Signals and Systems
Fundamentals of Electronics 3: Discrete-time Signals and Systems, and Quantized Level Systems

2017

MURET Pierre
Fundamentals of Electronics 1: Electronic Components and Elementary Functions

BUCCI Davide
Analog Electronics for Measuring Systems

2016

BAUDRAND Henri, TITAOUINE Mohammed, RAVEU Nathalie
The Wave Concept in Electromagnetism and Circuits: Theory and Applications

FANET Hervé
Ultra Low Power Electronics and Adiabatic Solutions

NDJOUNTCHE Tertulien
Digital Electronics 1: Combinational Logic Circuits
Digital Electronics 2: Sequential and Arithmetic Logic Circuits
Digital Electronics 3: Finite-state Machines

2015

DURAFFOURG Laurent, ARCAMONE Julien
Nanoelectromechanical Systems

2014

APPRIOU Alain
Uncertainty Theories and Multisensor Data Fusion

CONSONNI Vincent, FEUILLET Guy
Wide Band Gap Semiconductor Nanowires 1: Low-Dimensionality Effects and Growth
Wide Band Gap Semiconductor Nanowires 2: Heterostructures and Optoelectronic Devices

GAUTIER Jean-Luc
Design of Microwave Active Devices

LACAZE Pierre Camille, LACROIX Jean-Christophe
Non-volatile Memories

TEMPLIER François
OLED Microdisplays: Technology and Applications

THOMAS Jean-Hugh, YAAKOUBI Nourdin
New Sensors and Processing Chain

2013

COSTA François, GAUTIER Cyrille, LABOURE Eric, REVOL Bertrand
Electromagnetic Compatibility in Power Electronics

KORDON Fabrice, HUGUES Jérôme, CANALS Agusti, DOHET Alain
Embedded Systems: Analysis and Modeling with SysML, UML and AADL

LE TIEC Yannick
Chemistry in Microelectronics

2012

BECHERRAWY Tamer
Electromagnetism: Maxwell Equations, Wave Propagation and Emission

LALAUZE René
Chemical Sensors and Biosensors

LE MENN Marc
Instrumentation and Metrology in Oceanography

SAGUET Pierre
Numerical Analysis in Electromagnetics: The TLM Method

2011

ALGANI Catherine, RUMELHARD Christian, BILLABERT Anne-Laure
Microwaves Photonic Links: Components and Circuits

BAUDRANT Annie
Silicon Technologies: Ion Implantation and Thermal Treatment

DEFAY Emmanuel
Integration of Ferroelectric and Piezoelectric Thin Films: Concepts and Applications for Microsystems

DEFAY Emmanuel
Ferroelectric Dielectrics Integrated on Silicon

BESNIER Philippe, DÉMOULIN Bernard
Electromagnetic Reverberation Chambers

LANDIS Stefan
Nano-lithography

2010

LANDIS Stefan
Lithography

PIETTE Bernard
VHF / UHF Filters and Multicouplers

2009

DE SALVO Barbara
Silicon Non-volatile Memories / Paths of Innovation

DECOSTER Didier, HARARI Joseph
Optoelectronic Sensors

FABRY Pierre, FOULETIER Jacques
Chemical and Biological Microsensors / Applications in Fluid Media

GAUTIER Jacques
Physics and Operation of Silicon Devices in Integrated Circuits

MOLITON André
Solid-State Physics for Electronics

PERRET Robert
Power Electronics Semiconductor Devices

SAGUET Pierre
Passive RF Integrated Circuits

2008

CHARRUAU Stéphane
Electromagnetism and Interconnections

2007

RIPKA Pavel, TIPEK Alois
Modern Sensors Handbook

Printed and bound by CPI Group (UK) Ltd, Croydon, CR0 4YY

13/09/2026

14961037-0003